Biorhythms and Industrial Safety, Thumann, A. 166 p. The Fairmont Press. Atlanta, Georgia, 1977.

Reviewed by Henry M. Taylor, Chester, Virginia

This book presents the state-of-the-art opinions and scientific studies on biorhythms. It aims to provide a foundation for establishing experimental biorhythm programs within industry. Despite several typo errors, it is the seemingly bias by the author for biorhythms that encourages reader sceptacism. This reviewer came away with a feeling that this is an art of questionable merit looking for science. If this is so, nothing will be gained by application through an already overworked safety practioneer other than to increase book sales. As for application in general safety; much of the scientific information presented in the book precludes biorhythms as a sole viable tool for accident prevention.

BIORHYTHMS AND INDUSTRIAL SAFETY

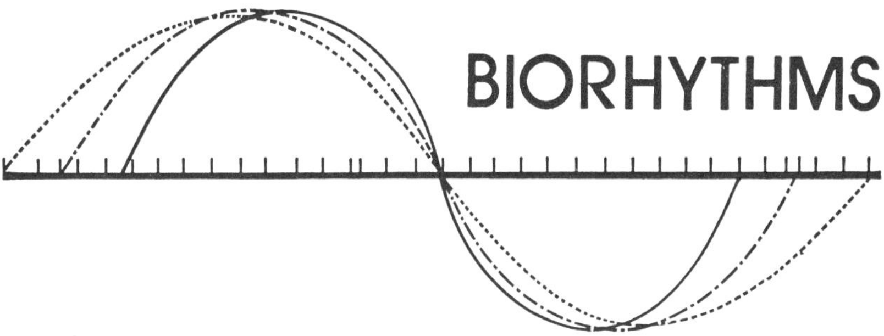

BIORHYTHMS AND INDUSTRIAL SAFETY

ALBERT THUMANN, P.E.

THE FAIRMONT PRESS
464 ARMOUR CIRCLE, N.E. • ATLANTA, GEORGIA 30324

Biorhythms and Industrial Safety

Library of Congress Catalogue Card Number 76-22974

ISBN: 0915586045

A JOB SIMULATION EXPERIENCE®

ACKNOWLEDGEMENTS

Appreciation is expressed to Mr. Jake Sanhein, Dr. Douglas Neil, Mr. Louis J. Giannetti, Mr. Thomas A. Wyatt, Mr. Francis L. Sink, Major John Wolcott, Lt. Col. Robert R. McMeekin, Mr. Robert E. Burgin and Mr. Robert E. Yanowitch who graciously provided the results of their research studies for this text. Special appreciation is given to Mr. Sanhein for the use of the Biorhythm Charts, which were specially prepared by him for use in this book.

Contents

1

Biorhythms and Industrial Safety

INDUSTRIAL ACCIDENT PREVENTION

With over four million industrial accidents recorded in 1975 alone, and one out of every ten workers afflicted with a job-related illness or injury,[1] it is evident why safety is one of industry's highest priorities. In 1970 the government established the comprehensive safety regulation, the Occupational Safety and Health Act (OSHA), the purpose of which was to raise the safety standards of the American industrial working environment. Even with this regulation industry safety records have not improved, and the majority of industrial accidents are attributed to individual worker performance. Traditional methods to reduce human errors have relied on educational programs, safety slogans, posters and verbal communications. This traditional approach assumes a constant level of human performance. Safety engineers usually do not take into account that human performance is a variable which is often not a function of the working environment. If emphasis can be placed on safety during low periods of human performance, this awareness itself may contribute to accident prevention. The science of "biorhythms" is the science of predicting human performance by means of understanding biological rhythms. By utilizing this new science, it is estimated that as many as two million on-the-job accidents might potentially be prevented each year.

WHAT IS THE SCIENCE OF BIORHYTHMS?

The word "biorhythm" is derived from the Greek words, *bios,* meaning life, and *rhythmos,* meaning a regulated beat. The biorhythm theory states that at the moment of birth, three cycles are initiated and recur consistently throughout a person's life. The three cycles have independent durations and influence the physical, emotional and intellectual states of the individual. Specific points in the cycle correspond to highs and lows in human performance.

THE DEVELOPMENT OF THE SCIENCE OF BIORHYTHMS

The science of biorhythms dates back to the turn of the century. In order to place biorhythm science in its proper perspective, a brief history of its development is appropriate.

INITIAL RESEARCH ABROAD

Dr. Herman Swoboda (1873-1963), a professor of psychology at the University of Vienna, did his initial research into the basic rhythms of man during the period of 1897 to 1902.[2] Dr. Swoboda's research included the study of the birth of infants among his patients and the documentation of hundreds of family trees. From this analysis, he developed a mathematical analysis of the 23 day physical cycle and the 28 day emotional cycle in man. Swoboda summarized the results of his research by stating that man's actions are influenced by periodic changes and that his reactions may be predicted. He showed that the psychologist can scientifically predict the periodic changes in man by use of a mathematical model. Swoboda also designed a slide rule whereby the "critical" days in a person's life could be computed based on the date of birth. Dr. Swoboda published several books on the subject, including *Die Perioden des Menschlinchin Lebens*[2] (The Periodicity in Man's Life), *Studien Zur Grundlegung der Psychologie*[3] (Studies of the Basis of Psychology), and *Das Siebenjahr*[5] (The Year of Seven).

Dr. Wilheim Fliess (1859-1928), a practicing physician in Berlin, published the findings of his research on fundamental rhythms in man during the period of 1895 to 1905.[7] Fliess' initial research dealt with a child's early immunity to a contageous disease subsequent to exposure to the disease. He traced the illness, the outbreak of the fever and deaths, and correlated them to the date of birth. He found that the child succumbed to the disease on a periodic day, and Fliess theorized that the 23 day and 28 day rhythms were fundamental to life. Fliess' analysis included complicated mathematical calculations and statistics which tended to confuse the medical profession. Dr. Fliess concluded, however, that the 23 day rhythm was related to the human physical condition, while the 28 day cycle was related to the nervous system and influenced a person's emotional well-being and his degree of sensitivity. Dr. Fliess published several books on the subject, including *Der Ablauf Des Lebens*[7] (The Course of Life), *Vom Leben und Vom Tod*[8] (Of Life and Death), *Das Jahr im Lebendigen*[9] (The Year of the Living), and *Zur Periodenlehre*[10] (The Theory of Periodicity).

Dr. Fliess' research was aided by the mathematician and medical doctor, Han Schlieper, who also published two books on the subject entitled, *Der Rhythmus im Lebendigen*[11] (Periodicity of Life), and *Das Jahr in Raum*[12] (The Year in Space).

Remarkably, Swoboda and Fliess developed their similar theories concerning the 23 and 28 day rhythms in man independently. Both researchers studied family trees and applied their knowledge of mathematics, psychology, and medicine to reach the conclusion that there is a biological clock in man.

THE DISCOVERY AT INNSBRUCK

Alfred Teltscher, a doctor of engineering and a teacher at Innsbruck, performed important research during the 1920's into the cyclic characteristics of the mind.[13] He found that his students' performance fluctuated in definite 33 day cycles. Teltscher believed that there were certain periods when a student could readily learn new subjects and other periods when the capacity to learn diminished. This periodicity was believed to have been caused by periodic, rhythmic secretions of glands affecting brain cells.

The existence of a mental cycle was also theorized by researchers in the United States during this same period. Dr. Rexford Hersey at the University of Pennsylvania, assisted by Dr. Michael Bennett, determined the presence of a 33 to 36 day rhythm related to mental function.[13]

BIORHYTHM CHARTING SIMPLIFICATIONS

Limitations in the application of the initial research by Fliess arose due to the confusing mathematics and charts associated with his works. During the late 1920's, Alfred Judt of Bremen, Germany designed calculation tables to correlate the day, month and year of birth with the date of an event in question. These tables were later refined by Hans R. Frueh, a Swiss engineer and mathematician.[14,15] Present day charting methods are based largely on these tables designed by Judt and Frueh and the slide rule developed by Dr. Swoboda in the early 1900's.

THE BEGINNING OF BIORHYTHM ACCEPTANCE

The science of biorhythms initially met with misunderstanding and skepticism within both the professional and private communities. The charting simplifications made investigation and analysis much easier and laid the groundwork for further development of the science.

SCHWING'S STUDY

In 1939 Hans Schwing, a student at the Swiss Federal Institute of Technology in Zurich, Switzerland, initiated a study which correlated accidents and death statistics with biorhythm cycles.[20] The basis of the study was the data of 700 accident cases and records of 300 deaths.

Schwing determined that a "critical day" occurs on days in which a new cycle begins or in which a cycle is changing from a high phase to a low phase. "Double critical" or "triple critical" days occur when this phenomenon occurs for two or three of the biorhythm cycles on the same day. On the 21,252nd day of a person's life (that is 58 years, 68 days), all three cycles will coincide precisely at a time corresponding to the time of day of the person's birth. During this period of 58 years and 68 days, a person will experience 16,925 non-critical days and 4,327 critical days.

Schwing's first analysis was made by correlating the accident data with the 23 day, 28 day, and 33 day cycles. Of the 700 accidents studied, 322 occurred on a single critical day, 74 on a double critical day, and 5 on a triple critical day. The remaining 299 accidents occurred on non-critical days. Schwing's results are summarized as follows:

Total Number of Accidents	700
Total Number of Accidents Occurring on Critical Days	401
Percentage of Accidents Occurring on Critical Days	60%
Percentage of Critical Days in a 58 Year, 68 Day Period	20%

These results showed that accidents occurred ***three times*** more frequently on critical days than would be expected by random occurrence, thereby indicating that man experiences days during which his coordination is impaired sufficiently to increase the probability of human error. Since Schwing chose the 700 accident cases at random, the sample contained some accidents which did not result from individual human performance errors. If the sample had been selected to include only accidents which were self-induced, the correlation with critical days might be expected to be even greater.

Schwing made a second analysis based on the 300 death records. In this case he used only the 23 day and 28 day (physical and emotional) cycles. Of the 300 deaths, 197, or 65% occurred on critical days. The percentage of critical days which would randomly occur is only 15 percent, showing a very positive correlation for this sample for the occurrence of death on biorhythm critical days.

THE WORKS OF FRUEH

Another researcher whose work contributed to the acceptance of biorhythm science was Hans R. Frueh. In 1946 Frueh published two books entitled, *Wie Berechnet und Werted man ein Rhythmogram?*[16] (How are Biorhythm Charts Calculated and Evaluated?) and *Kraft, Gesundheit und Leistung*[17] (Strength, Health and Performance). Frueh's work encouraged medical doctors in Switzerland and Germany to use biorhythm calculations for surgery patients. Also at this time, safety engineers in the transportation industry began to employ biorhythms for accident prevention.

MODERN BIORHYTHM DEVELOPMENT

Two major biorhythm studies relating to industrial accidents were published in the 1950's. In 1954 Reinhold Bochow, under the direction of Dr. J. Sennewald at the Humboldt University in Berlin, reported on an investigation involving accidents of agricultural machinery workers.[21] In the analysis of 497 accidents it was found that 97.8 percent of the accidents fell on critical days. It was Bochow's theory that everyday actions become automatic and are effected on a subconscious level. When rhythmical influences disturb the subconscious, an individual becomes more prone to error, and consequently to error-induced accidents.

In another study, also during the 1950's, O. Tope, Chief Engineer of the Department of Sanitation of the City of Hanover, West Germany, prepared a report on accidents suffered by shop workers, street cleaners, and truck drivers.[22] This report concluded that 83 percent of the accidents occurred on biorhythm critical days.

DEVELOPMENT IN THE UNITED STATES

Although the science of biorhythms did not have its roots in the United States, numerous studies have been performed within this country. In fact, at least 100 organizations have experimented with biorhythms, but unfortunately the results of many of these studies have remained confidential due to the widespread controversy surrounding the topic. It is not surprising then that the science of biorhythms has not been applied extensively within American industry to date.

The efforts of George S. Thommen, a native of Switzerland, have contributed largely to increased interest in and application of biorhythms in the United States as well as abroad. Mr. Thommen is the author of numerous articles and research reports and has spent over twenty-five years probing the original discoveries relating to biorhythms in Austria, Germany and Switzerland. In 1973 he published the widely read book, *Is This Your Day?*[13] and has also prepared a text for an IBM 360 computer program for biorhythm analysis. Thommen has studied the relative influence of each of the three biorhythm cycles, and his research has indicated that the 33 day intellectual cycle has only a minor influence on human error in contrast to the physical and emotional cycles.

Thommen's work has also had a major influence on other investigators. In the early 1960's Russell K. Anderson analyzed more than 1,000 accidents.[23] His results showed that 90% of the accidents had occurred on biorhythm critical days. Anderson pointed out several potential sources of error in biorhythm analysis, such as the fact that non-English speaking employees do not always give accurate dates of birth. He recommended that biorhythm analysis be performed for all employees who are involved in operations considered hazardous and in all areas where there is a possibility of far-reaching and serious accidents. Industries involved in the chemical or nuclear fields would be prime examples. His method for using biorhythms in accident prevention was presented at the 35th annual Eastern Regional Safety Convention in April 1965.

Anderson emphasized that the key to using biorhythms is to realize that they are not a predictor of accidents, but rather an indicator of the biological stresses which affect an individual. Thus, when a person is aware he is at a biorhythm "low," he can take extra care to avoid errors which might result in accidents. The procedure for applying biorhythms to industrial safety programs, then, simply consists of determining employees' critical days and then making this information available to the individual employees, as well as to the supervisor, who may utilize it for teaming employees. This procedure has been shown to reduce accidents in actual practice.

THE RESEARCH OF DR. NEIL

In 1974 Dr. Douglas E. Neil of the Naval Post Graduate School in Monterey, California published the results of several biorhythm investigations.[24] It was Dr. Neil's feeling that examining only "critical days" results in a partial analysis, and regardless of outcome, the results can only be interpreted as inconclusive. Dr. Neil investigated the relationship between accidents and biorhythms by randomly selecting reports of 127 industrial accidents which occurred at an aircraft maintenance facility. The results indicated that the distribution of accidents was not uniform, and a statistically significant correlation was observed for the physical and intellectual cycles. Recognizing a relationship between accidents and biorhythm cycles, Dr. Neil then proceeded to develop experiments to measure human performance under stringent laboratory conditions. Experimental subjects were asked to perform information processing tasks on a daily basis for a period of 70 days. The tasks varied in level of difficulty, and performance was measured in terms of information processing rates, reaction times, and movement times. Extraneous variables were controlled. It was found that individual performance varied significantly on different days. The data was then analyzed to determine periodicity. A Fast Fourier Transform Analysis was performed by which frequency envelopes and amplitude distributions of harmonics were determined. The computer at the Naval Post Graduate School was then used to find the phase of each harmonic and the significance of the frequency data. It was determined that of the fourteen periods detected in the frequency envelopes, nine periods were within one day of the critical zones of the three biorhythm cycles. For the 23, 28, and 33 day cycles, the periods are 11.5, 14, and 16.5 respectively. All subjects had significant periods within one day of these values. The probability of these results occurring by chance is 1.5 in 1,000.

The research conducted at the Man-Machine Systems Design Laboratory at the Naval Post Graduate School lead to the conclusion that a periodicity exists in human performance which approximates the hypothesized biorhythm oscillation. Dr. Neil concludes his report by stating that human performance is the result of interaction of a multitude of factors of which biorhythms are but one. Hence, a program which is directed toward reducing human variability for the purpose of reducing accidents or increasing productivity must not restrict itself to the consideration of any single factor, including biorhythm analysis.

WHY THE CONTROVERSY OVER BIORHYTHMS?

Not all professional opinions concur with regard to the validity of the biorhythm theory. However few of the books or publications on the subject to date have recognized the many scientific studies and reports which deny the relationship between biorhythm cycles and human performance. This might be attributed to the fact that many of the promoters of biorhythms have a vested interest in the success of this science. In order to present an objective viewpoint concerning biorhythms, some case studies which contest the validity of the science will be discussed.

In 1972 the biorhythmic characteristics of 1,051 records of sporting events were analyzed by the Biorhythm Research Center in Oberwil, Switzerland.[25] This investigation concluded that there is no statistical method to prove any influence of biorhythms based on the results of such events. This study concluded that biorhythms are based on a theory without biological or mathematical foundations. The invalidity of biorhythms was demonstrated by accepted statistical methods and by using a sufficiently large collection of suitable dates.

In November 1975 Major John Wolcott and Lt. Col. Robert McMeskin of the Armed Forces Institute of Pathology along with Robert Burgin of the National Transportation Safety Board and Robert Yanowitch of the Federal Aviation Agency published a study relating to aviation accidents.[26] A total of 4,279 aircraft accidents which were found to be "pilot involved" were analyzed with regard to correlation with biorhythm critical days. It was concluded based on the results of the study that a causal relationship between biorhythm critical days and aircraft accidents could not be established. (A reprint of the results of this study may be found in Chapter 4.)

In the early 1970's Keith Mason studied 13,285 accident records from the Workmen's Compensation Board of British Columbia.[27] He concluded that accidents are no more likely to occur during critical periods than at any other time. His initial report included all accidents, including those which were not entirely the fault of the victim. Mason subsequently analyzed a subsample of 5,259 accidents which were determined more likely to be the fault of the victim. These included accidents such as falls. The results were the same: no significant correlation was found between the occurrence of the accidents and biorhythm critical days.

Kenneth Tyler of the Truck Underwriter's Association of Los Angeles conducted another study of 18,000 insurance group policies.[28] The initial results indicated that 50 percent of the accidents occurred on biorhythm critical days. The results were further analyzed by the firm's statistical department. The statistical analysis showed the pattern of accidents to be the same as might be expected from chance occurrence.

Some of the reasons for the differing conclusions of the various investigators include the following:

1. The methods used for correlating human behavior and biorhythms does not always conform with recognized statistical methodology. A single event, selectively chosen, cannot statistically prove or disprove a hypothesis. Many books have used the experiences of individual celebrities to illustrate the principles of biorhythms. Although these may provide interesting examples, they do little to scientifically demonstrate the validity of biorhythm theory.
2. Inaccuracies in recording birthdates or in the methods used for calculating biorhythm cycles may lead to false conclusions.
3. Many reports on accident prevention indicate that when employees are made aware of their critical days, accidents are reduced. There is some question, however, as to whether this accident reduction is a result of using biorhythms. If employees were made aware of their performance on a random basis, would this in itself not reduce accidents?
4. Few studies dealing with biorhythms provide back-up data and outline the method used for evaluation. Most researchers have not made available their assumptions relating to the hour of birth or the critical day or period (24 or 48 hours) used for their calculations. The definition of the critical day is important. For example, does it begin at midnight or noon? How does the researcher account for the 23 and 33 day cycles, which change phase at half days?

5. Few studies have been presented on the same basis. No standards have been developed for items such as:
   a. Basis for inclusion of accidents. (Which accidents have been excluded from the analysis has an important effect on the conclusions reached.)
   b. Definition of terms.
   c. Methods for statistical verification.
6. Sensational reports, exaggerated correlations, reports on the effect on one's sex life, and the like have tended to destroy the credibility of biorhythms as a science.
7. Scientific investigations have shown that there are "clocks" in man other than the biorhythm cycles. It is not clear how these other body time cycles might influence human performance and thereby affect the results of biorhythm studies.

JAPANESE INDUSTRIAL APPLICATIONS

In Japan biorhythm programs have been applied in several industries. Mr. Yujiro Shirai, a representative for the Japan Biorhythm Association, reported the following:[13]

1. Biorhythm charting is used by many companies. These include: Hitachi's Manufacturing Company, Fuji Heavy Industries, Asahi Glass, Mitsubishi Heavy Industry, Bridgestone Tire, Tokyo Electric Company, Chubu Electric Company, Tohoku Electric Company, Tokyo Marine & Fire, Yasuda Fire, Taisho Fire, Japan Fire Insurance Company, Mitsui Life, Meiji Life, Asahi Life, Fukoku Life, Japan Express, Seino Transport, Yamato Transport, International Automobile, and Odakyo Taxi Cab Companies.
2. The Nagahama Transportation division of the Omi Railroad uses biorhythm charting and reported an accident-free record covering over four million kilometers.

3. The Meigi Bread Company uses biorhythm charting and reported that it reduced its yearly vehicle accident loss rate by 45 percent.
4. The Shigi and Gumma Perfecture Traffic Safety Associations are now issuing monthly biorhythm charts to 250,000 drivers.

The apparent successful use of biorhythm charting in Japanese industrial safety programs substantiates the need for validating the applicability to safety programs of the biorhythm theory for American industry. Increased experimental and practical use of biorhythms should serve to resolve much of the controversy which has resulted from the scientific investigations.

THE STATUS OF BIORHYTHM APPLICATIONS IN AMERICAN INDUSTRY

There have been many claims as to the widespread use of biorhythms in American industry. *National Safety News* conducted a survey[28] to substantiate claims by biorhythm proponents that the theory is widely used in occupational safety programs. Biorhythms are a controversial subject, and positions of companies in this area are in constant flux. It should be noted that the data summarized here from the *National Safety News* survey has not been confirmed by the author, nor should it be interpreted as a firm position of the companies listed. The following is a summary of the response of each company contacted as reported by *National Safety News:*

National Lead. National Lead experimented with biorhythms from 1969 to 1971. The program was discontinued as a result of loss of access to a computer. The two-year experiment was conducted by Mr. Lawrence Newcomb at the National Lead New Jersey plant. At the time of the experiment the accident rate had been consistently rising within the plant. The biorhythm program was conducted using one experimental group and two control groups. For Group I, the experimental group, biorhythm critical days were calculated, and the information was given to the foreman to review with his workers. For Group II, the first control group, critical days were calculated but incorrectly reported to the foreman and workers. This was done in order to determine whether the extra attention resulting from the biorhythm program was itself a significant factor in accident prevention. For Group III, the second control group, no critical day calculations were made or reported to workers. The results: Group I, no increase; Group II, the accident rate increased slightly; and within Group III it increased substantially. This study showed the use of biorhythms as well as extra attention to safety to be factors in accident reduction within the plant.

United Airlines. United Airlines used biorhythms as part of a safety program at its San Francisco maintenance facility. The company recently discontinued the program and issued the following official statement:

> The Biorhythm program, which was utilized to a certain extent in United Airlines, is not and never has been used to schedule work assignments for employees. It definitely has not been used to determine if pilots should fly, because we have a very specific system of determining if the pilot is fit to fly through proficiency and line checks and other ways approved by the Federal Aviation Administration.
>
> The Biorhythm program was used during the last couple of years primarily as a tool to create a greater awareness of safety. We do not have any hard data that will indicate that the Biorhythm program had any direct effect on our ground safety performance, and the program has been discontinued.

Allegheny Airlines. Allegheny Airlines is experimenting with biorhythms but has released no details of their results.

Swissair. This airline reported that they have never used biorhythms.

Zurich Municipal Transit Company. This company also reported never to have used a biorhythm program.

AT&T. AT&T Longlines is experimenting with biorhythms at a local New York office but has not yet reported on results.

Pfizer, Inc. Pfizer has been experimenting with biorhythms for twelve years. Biorhythm calculations have been used for rotation of production workers, resulting in a 60 percent drop in accidents. In a study conducted by the company involving 1,000 accidents, it was found that 30 percent of the accidents had occurred on critical days.

Exxon Chemical Company. Exxon Chemical, Daytown, Texas is commencing a biorhythm program involving 900 employees. Each employee will be sent a safety reminder on each of his critical days.

Naval Air Development Center. At the Naval Air Development Center, Warminster, Pennsylvania, Lt. Commander Arner is starting a program in which he distributes biorhythm charts to his pilots along with information which includes criticism of biorhythm theory. The program is intended simply to arouse interest.

Transport Indemnity. This company reported plans to initiate a biorhythm safety program for one of the trucking companies which it insures.

Zidell Explorations. This building firm in Portland, Oregon has examined the records of a small number of past accidents and found that about two-thirds of the accidents seemed to coincide with calculated biorhythm critical periods. The company plans to continue its investigation.

Gulf Oil of Canada. Gulf Oil of Canada has developed a computer program to determine correlation between accidents and biorhythm theory. If a significant correlation is found, a biorhythm safety program will be implemented.

Pacific Northwest Bell. A proposed plan to provide biorhythm charts to Pacific Northwest Bell's 6,000 employees was recently rejected in spite of a study which showed a correlation between accidents and critical days.

American Can Company. An American Can Company plant in Naheola, Alabama has determined a reasonably good correlation between accidents and employee critical days. A biorhythm safety program has not been implemented as yet, however, due to lack of confidence in a positive worker reaction to the program.

In general the *National Safety News* study found that many companies were reluctant to discuss their biorhythm programs. Apparently there is a general feeling that knowledge of their biorhythm applications might damage their public image. Another area of concern is the potential opposition or other problems from labor unions. B. J. Krause-Poray tried to introduce biorhythms for industrial safety programs in Australia. He found opposition from unions, who claimed that biorhythm programs would invade the worker's privacy. In the United States, Russell Anderson, a retired safety consultant, found that in some cases unions expressed the opinion that workers should receive holiday pay when they work on biorhythm critical days.[28]

As the above study indicates, American industry seems at present to combine interest with due skepticism of industrial safety applications of biorhythm theory. It is evident that experimenting is continuing, however, and if positive results continue to be achieved and made known, it is likely that biorhythm programs will become an accepted part of industrial safety procedures.

OTHER BIOLOGICAL RHYTHMS

Cyclic fluctuations in human biological function other than the 23, 28 and 33 day biorhythm periods have been observed and studied with regard to their effects on performance. Within a single 24-hour period, regular periodic variations in body temperature, blood pressure, respiration, pulse, blood sugar, hormone levels, urine, cell division, hemoglobin levels and amino acid levels have been shown to occur.[29] Rhythms associated with these variations are referred to as "circadian" rhythms (from Latin *circa* meaning "near" or "around" and *dies* meaning day).

Dr. K. E. Klein and a group of researchers at the Institute for Flight Medicine in Bad Godesberg, West Germany have conducted experiments to measure daily variations in human performance.[29,30] The researchers measured several body functions along with reaction time in a series of complex ten-minute tests. Based on the study it was reported that the subjects' mental performance was optimum between 2 and 4 p.m. and at its lowest level between 2 and 4 a.m.

In another study conducted by the German Flight Institute, it was concluded that best mental performance occurs between 1 and 7 p.m., while the worst occurs between 2 and 6 a.m.[29]

Some studies have shown that industrial accidents occur with greatest frequency between 2 and 4 a.m., although other studies have reached differing conclusions. In Sanhein's study,[31] which is presented in Chapter 3, it was also concluded that the probability of accidents is greatest in the late evening hours.

WORK SHIFTS

Dr. Ahlgren of the University of Minnesota has studied the effects of variations in the daily work schedules of workers.[28] He reported that the body functions of an individual are temporarily impaired by a significant change in the working schedule, and until the worker's body adjusts to the new hours, the likelihood of accidents is increased. It has been found that a reversal of an individual's sleeping and waking hours may be followed by a biological transition or adjustment period as long as three weeks.

There is increasing evidence that rotating shifts is detrimental to health. Ulcers and hypertension are common health problems among air traffic controllers, who frequently rotate shifts. Another interesting finding is that some people are able to adjust to phase shifts more easily than others. Thus, it may be possible to pre-screen individuals for shift work. Only those found to have inherent adaptability to phase shifts could be scheduled to perform on this basis.[28]

OTHER STUDIES OF CIRCADIAN RHYTHMS

An experiment was performed in which a group of men were placed on an eighteen rather than twenty-four hour day.[29] It was found that the men had increased vulnerability to certain types of infections, and that they became restless and emotionally tense. It was also observed that the tasks performed by these men were impaired by lack of coordination.

Studies such as these related to circadian rhythms serve to demonstrate that there are indeed many time-related biological factors which influence human performance. Continuing investigations leading to a better understanding of the interrelationship of all the biological "clocks" in man affecting his performance will hopefully, among other benefits, lead to more effective methods of accident prevention.

ASTROLOGY AND BIORHYTHMS

The science of biorhythms is often incorrectly associated with astrology. It is important to recognize that these two fields are separate areas of study and neither contradict nor support one another.

ORGANIZATION OF THIS BOOK

This book is designed to illustrate the state-of-the-art opinions and scientific studies on biorhythms and to provide a foundation for establishing an experimental biorhythm program for your industrial facility.

In Chapter 2 you will learn the basic mathematical properties of biorhythms, including modeling, probability and statistics, and a definition of accident-prone days.

In Chapter 3 you will see how Jake Sanhein applied the biorhythm theory to show a relationship between 1,300 accidents examined and biorhythm cycles.

Chapter 4 illustrates another recognized authority's work relating to biorhythms and aircraft accidents. In Major John Wolcott's study you will see a case in which no relationship could be established between biorhythms and the accidents studied.

Chapter 5 discusses how biorhythms are thought to influence human behavior.

A statistical experiment performed by Dr. Douglas Neil which supports biorhythms' influence on human performance is presented in Chapter 6. Chapter 7 presents the results of a controlled laboratory experiment conducted by Dr. Neil which also shows a relationship between the biorhythm cycles and human performance.

Chapter 8 demonstrates how the state-of-the-art theories may be used to perform an in-plant biorhythm test program for an industry. This chapter contains simplified tables which will assist you in examining the accident data of your plant without a computer or time-consuming plotting.

REFERENCES

1. *Accident Facts,* National Safety Council, Chicago, 1976 Edition.
2. Swoboda, H., *Die Perioden des Menschilichen Lebens,* Franz Deuticke, Leipzig and Vienna, 1904.
3. Swoboda, H., *Studien zur Grundlegung der Psychologie,* Franz Deuticke, Leipzig and Vienna, 1905.
4. Swoboda, H., *Die Kritischen Tage des Menschen,* Franz Deuticke, Leipzig and Vienna, 1909.

5. Swoboda, H., *Das Siebenjahr,* Orion-Verlag, Vienna and Leipzig, 1917.
6. Swoboda, H., *Die Bedeutung des Siebenjahr Rhythmus fuer die Menschliche Vererbung,* Industria Tipografica Fiorentina, Firenze, 1954.
7. Fliess, W., *Der Ablauf des Lebens,* Franz Deuticke, Leipzig and Vienna, 1906.
8. Fliess, W., *Von Leben und vom Tod,* Eugene Diederichs, Jena, 1909.
9. Fliess, W., *Das Jahr im Lebendigen,* Eugene Diederichs, Jena, 1918.
10. Fliess, W., *Zur Periodenlehre,* Ebenda Verlag, Leipzig, 1925.
11. Schlieper, H., *Der Rhythmus des Lebendigen,* Eugene Diederichs, Jena, 1909.
12. Schlieper, H., *Das Jahr im Raum,* Eugene Diederichs, Jena
13. Thommen, George S., *Is This Your Day?,* Avon Books, New York, 1976.
14. Frueh, H., *Von der Periodenlehre zur Biorhythmenlehre,* Wegweiser-Verlag, Zurich and Leipzig, 1939-1942.
15. Frueh, H., *Rhythmenpraxis,* Verlag H. R. Frueh, Bassersdorf, Zurich, 1943.
16. Frueh, H., *Wie Berechnet und Werted man ein Rhythmogramn?,* Verlag H. R. Frueh, Bassersdorf, 1946.
17. Frueh, H., *Kraft, Gesundheit und Leistung.* Selbstverlag, Bassersdorf, 1946.
18. Frueh, H., *Deine Leistungskurve,* Verlag H. R. Frueh, Bassersdorf, Zurich, 1953.
19. Frueh, H., *Triumph der Lebensrhythmen,* Lebensweiser-Verlag, Buedingen-Gettenbach, 1954.
20. Schwing, H., *Ueber Biorhythmen und Deren Technische Anwendung,* Leeman & Co., 1939.
21. Bochow, R., *Der Unfall im Landwirtschaftlichen Betrieb,* Humboldt University in Berlin (dissertation), 1954.

22. Tope, O., *Biorhythmische Einfluesse und ihre Auswirkung in Fuhrparkbetrieben,* report prepared for Sanitation Department, City of Hanover, West Germany, September, 1956.
23. Anderson, R., "Biorhythm–Man's Timing Mechanism," *ASSE Journal,* February 1973.
24. Neil, D., "Biorhythms and Industrial Safety," *National Safety Congress Transactions,* Vol. 2, 1974.
25. Schonholzer, G., Schilling, G., and Muller, H., "Biorhythmics," from National Safety Council Synopsis 730455, 1972.
26. Wolcott, J. H., McMeekin, R. R., Burgin, R. E., and Yanowitch, R. E., "Biorhythms–Are They a Waste of Time?" *TAC ATTACK,* Vol. 15, No. 11, 1975.
27. Mason, K., "An Investigation of the Biorhythm Theory," Workmen's Compensation Board of British Columbia, 1971.
28. Hirsh, T., "Biorhythm–Or Is It a Critical Day?" *National Safety News,* February 1976.
29. Luce, G. G., *Body Time,* Bantam Books, New York, 1973.
30. Gittelson, B., *Biorhythm–A Personal Science,* Arco Publishing Co., New York, 1975.
31. Sanhein, J., "Biorhythm Analysis as Applicable to Safety," *National Safety Congress Transactions,* Vol. 14, 1975.
32. Newcomb, C., "Bio-Rhythm Chart Report," address to Edison Electric Institute Accident Prevention Committee, Milwaukee, Wisconsin, 28 October 1974.
33. Brady, T., "Bio-What?" *TAC ATTACK,* Vol. 15, No. 11, 1975.

2

Basics of Biorhythms

MODELING TECHNIQUES

The charting of the three biorhythm cycles in man is modeled after mathematical sinusoidal functions. Why should the sinusoid be selected as the model for these biological cycles? Indeed, nature in general seems to have a decidedly sinusoidal character; the motion of a pendulum, the natural response of an underdamped second-order electrical or mechanical system, the bouncing of a ball, the vibration of a violin string, the political atmosphere in any country, and the economy, will almost always display a reasonably sinusoidal character, as do the biorhythm cycles in man.

The basic rhythm, A, may be denoted by a sine wave with period T:

$$A = \sin\left[\frac{2\pi}{T}(t + \phi)\right] \qquad (2\text{-}1)$$

where: $2\pi = 360°$
T = period in days
t = time in cycle days
ϕ = phase displacement

Figure 2.1 illustrates a typical biorhythm cycle at a specified point in time.

At the moment of birth each of the three biorhythm cycles is initiated, and the cycles proceed to follow a fixed sinusoidal pattern throughout the life of an individual. The physical cycle,

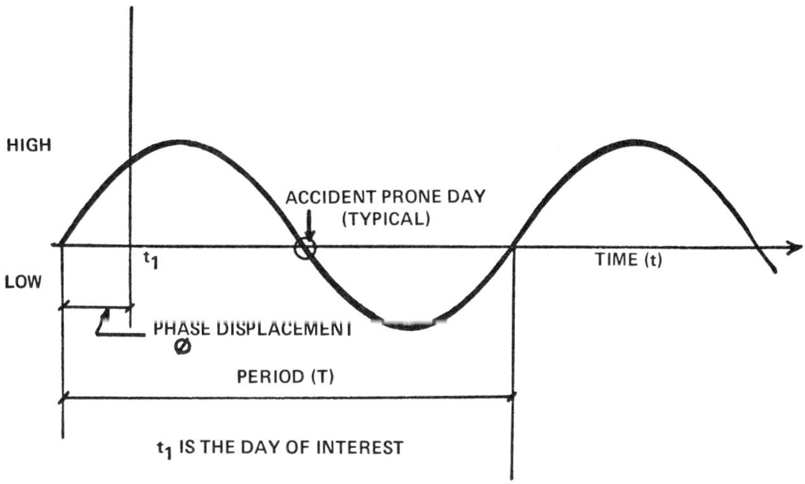

Figure 2.1 Typical Biorhythm Cycle

P, has a period of 23 days, the sensitivity (or emotional) cycle, *S,* a period of 28 days, and the intellectual cycle, *I,* a period of 33 days. Individuals with different birth dates will consequently have different composite biorhythm charts, although the theory holds that the cycles of all individuals follow the same 23, 28, and 33 day biological rhythm. The three biorhythm cycles may be described at any point in time by the following relationships:

$$P = \sin\left[\frac{2\pi}{23}(t + \phi_1)\right] \tag{2-2}$$

$$S = \sin\left[\frac{2\pi}{28}(t + \phi_2)\right] \tag{2-3}$$

$$I = \sin\left[\frac{2\pi}{33}(t + \phi_3)\right] \tag{2-4}$$

where: P = the physical cycle
ϕ_1 = physical phase displacement
S = the sensitivity cycle
ϕ_2 = sensitivity phase displacement
I = the intellectual cycle
ϕ_3 = intellectual phase displacement

Figure 2.2 illustrates the three biorhythm cycles.

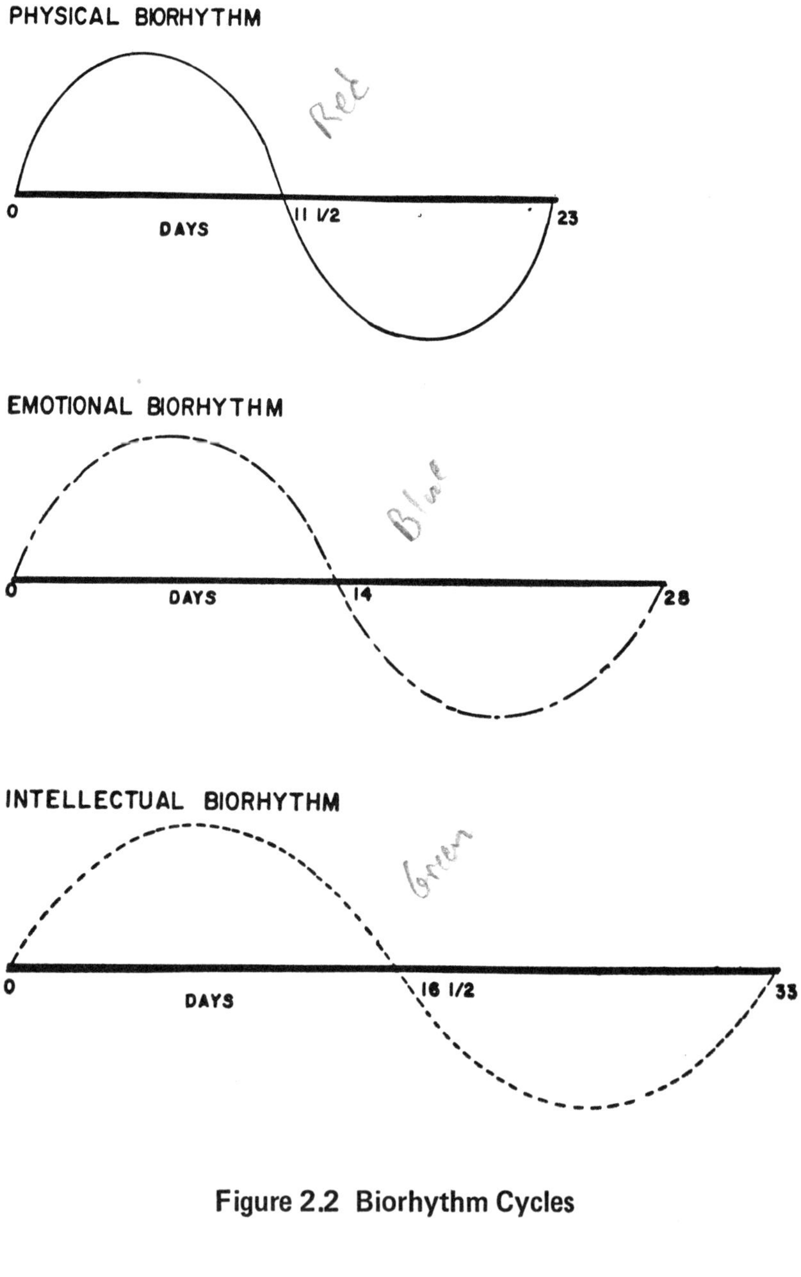

Figure 2.2 Biorhythm Cycles

The calculation of an individual's biorhythm at any given time requires first that the date being investigated be specified. The subject's age in days from the date of birth, up to and including the date of interest, must next be determined. In the calculation, consideration should be given to regular leap years and centurial leap years. The phase displacement may then be calculated and employed in Equations (2-2), (2-3), and 2-4):

$$\phi = \left(\frac{Y}{CL} - CC\right) \qquad (2\text{-}5)$$

where: ϕ = phase displacement, or day of the current biorhythm cycle on which the observed event occurred

Y = subject's age

CL = cycle length in days (23 for physical, 28 for sensitivity, 33 for intellectual)

CC = whole number of completed cycles (whole number associated with Y/CL)

INTERPRETATION OF CYCLES

In any of the three cycles, the positive period is thought to represent favorable conditions; that is, a high performance interval for intellectual function *(I)* or for physical coordination *(P)*, and a period of positive feelings, such as optimism and cheerfulness associated with the sensitivity cycle *(S)*. On the other hand, the negative period is thought to represent a "recharging" phase. During the recharging phase it is believed that a person is inclined to tire more easily *(P)*, become depressed or irritable more readily *(S)*, and exhibit a lesser degree of acuity in the learning and decision-making processes *(I)*.

ACCIDENT PRONE DAYS

"Accident prone days" are said to occur when one's energy changes from the positive phase to the negative phase or vice versa. These days, also called "critical days," are considered to be accident prone because the body's system is in a state of transition and is not stable. This instability does not in itself cause accidents, but does apparently have a mild negative influence on performance which may increase the probability of accident occurrence, particularly if a person is exposed to a situation of difficulty or danger. It is believed that an individual may compensate for this mild negative influence simply by being aware that he is experiencing an accident prone day. Accident prone days may be easily calculated in advance. The occurrence and characteristics of accident prone days are summarized in Table 2.1.

Table 2.1 Summary of Accident Prone Days

CYCLE	*ACCIDENT PRONE DAYS – 24 HOUR DURATION (1)*	
Physical	First Day @ 12 PM	Twelfth Day @ 12 AM
Sensitivity	First Day @ 12 PM	Fifteenth Day @ 12 PM
Intellectual	First Day @ 12 PM	Seventeenth Day @ 12 AM

(1) Period extends 24 hours beyond time specified.

COMBINED BIORHYTHM CYCLES

The three biorhythm cycles may be charted on one curve as indicated in Figure 2.3. The three rhythms are plotted independently, and their relative positions will change from month to month. The example in Figure 2.3 illustrates several accident prone days. Accident Prone Day 1 shows a crossing of the physical cycle. At this point the sensitivity cycle is recharging, while the intellectual cycle is at a high. On the other hand, Accident Prone Day 2 illustrates a double crossing of the zero axis, or a

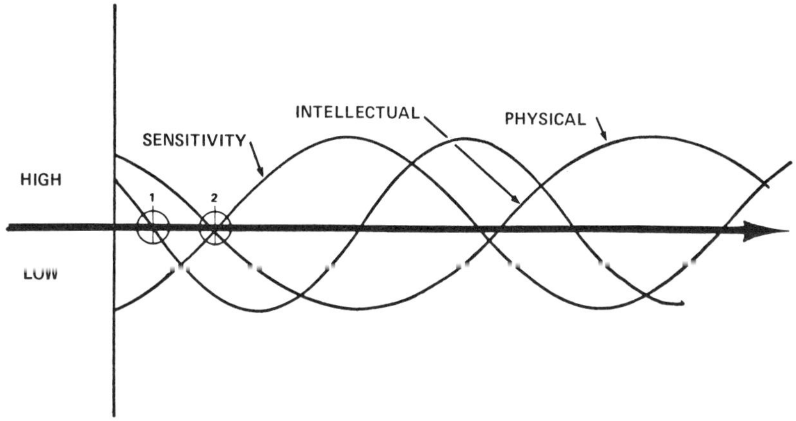

Figure 2.3 Combined Biorhythm Cycles

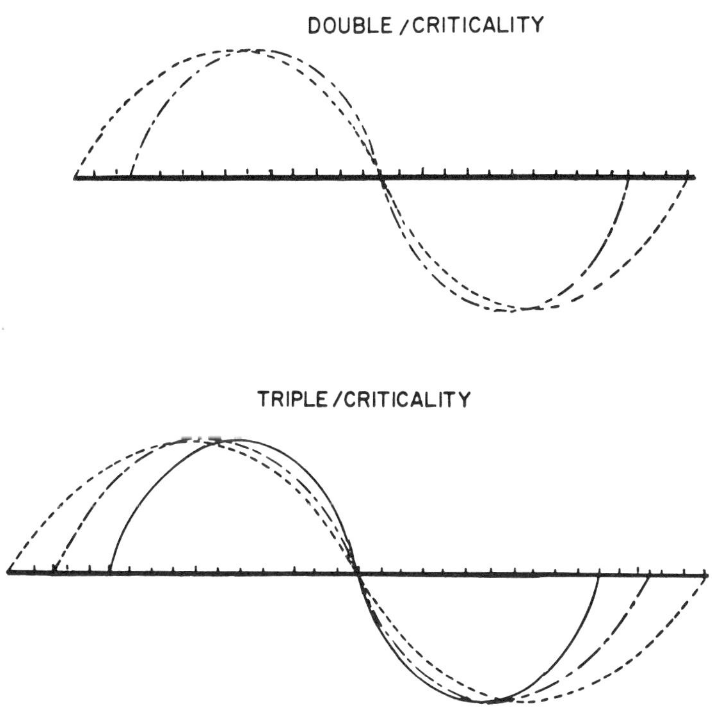

Figure 2.4 Double and Triple Critical Days

"double critical day." Similarly, three crossings within the same day result in a "triple critical day." Double and triple critical day occurrences are demonstrated in Figure 2.4.

FUNDAMENTALS OF PROBABILITY AND STATISTICS AS APPLIED TO BIORHYTHM TESTING

In order to scientifically demonstrate whether or not a relationship exists between biorhythms and human performance, it is necessary to have some knowledge of probability and statistics.

The probability of occurrence of an accident prone day, or a zero crossing of a biorhythm curve may easily be calculated. If all of the zero crossings of the three cycles in a year are added and divided by the number of days in a year, the calculated probability of 21.9% is found. This value is reduced to 20.4% by excluding multiple crossings on the same day.

The probability of a zero crossing of any single cycle or of multiple crossings is presented in Table 2.2. It is noted that the longer the cycle period, the higher the probability of a zero crossing.

Since the accident prone periods for the physical and intellectual cycles will alternately center on midnight and noon for adjacent cycles, it is often difficult to accurately assign an accident to these biorhythm periods. It is therefore convenient to assess a 48 hour period for alternate periods of the physical and intellectual cycles. The expected percentage occurrences of accident prone days for this analysis are presented in column three of Table 2.2. It is found in this case that the probability of occurrence of an accident prone day is 26.6%.

Each biorhythm accident prone period may also be analyzed as a 48 hour period. The fourth column of Table 2.2 shows the expected percentage occurrences of accident prone periods for a 48 hour period. The expected occurrence of these accident prone periods is 37.8%.

Table 2.2 Expected Percentage Occurrences of Accident Prone Periods

ACCIDENT PRONE DAY	*24 HOUR PERIOD*	*SEE NOTE*	*48 HOUR PERIOD*
Single Physical *(P)*	7.5850	11.0107	13.099
Single Sensitivity *(S)*	6.1265	5.6465	10.3707
Single Intellectual *(I)*	5.1383	7.3404	8.5827
Double *(P)* and *(S)*	.5835	.8470	2.1834
Double *(P)* and *(I)*	.4894	1.1011	1.8069
Double *(S)* and *(I)*	.3953	.5647	1.4305
Triple *(P)* and *(S)* and *(I)*	.0376	.0847	.3011
TOTAL ACCIDENT PRONE	20.3557	26.5951	37.7752
TOTAL NON-ACCIDENT PRONE	79.6443	73.4049	62.2248

NOTE: 24 hour period for accident prone day 1 based on all cycles, and day 15 for the sensitivity cycle. 48 hour period for half day crossings on day 12 for physical cycle and day 17 for intellectual cycle.

Finally, the zero crossing of a biorhythm cycle may be considered to indicate a 72 hour accident prone period. For a 72 hour period, the probability of an accident prone period occurrence is 51.5%.

In the chapters which follow, reference will be made to various statistical terms and tests. In the following sections, some of the fundamentals of statistics applicable to the study of biorhythms will be presented.

BASIC TERMINOLOGY

Estimated Mean ($\hat{X}$): An estimate of the true mean, usually based on the results of a sample or samples.

Normal Probability Curve: The symmetric lump-backed (bell-shaped) curve that closely approximates a large number of practical situations.

Normality Assumption: The assumption that the situation under consideration can be described as following an approximately normal curve.

Parameter: Any descriptive measure of a population (for example, the standard deviation and mean are parameters).

Population: Each and every member of some group (for example, all the individuals who had a lost time accident in a company).

Probability (P): The statistical likelihood that a given event or circumstance will occur. Probabilities range from 0 (least likelihood) to 1 (greatest likelihood).

Standard Deviation (S): A measure of the amount which individual values of a sample are dispersed about the mean. The larger the standard deviation, the farther the results of a particular sample can be expected to deviate from the true mean. The standard deviation of a sample is an estimator of the standard deviation (σ) of the entire population from which the sample is taken.

True Mean (μ): The arithmetic average or central tendency of a given characteristic parameter of a population which considers all members of the population.

CHI-SQUARE GOODNESS OF FIT TEST

Frequently it is necessary to determine whether the result of a biorhythm analysis might reasonably have been expected to occur purely by chance. The Goodness of Fit test is used to test the conformity of the sample to a hypothetical distribution which is defined by:

$$\lambda^2 = \sum \left(\frac{(n_i - P_i n)^2}{P_i n} \right) \qquad (2\text{-}6)$$

where: λ^2 = a measure of the discrepancy between the observed number of accidents and that predicted by pure chance.

n_i = the observed number of accidents in each category.

n = the total number of accidents.

P_i = the expected probability of each category (Table 2.2).

$P_i n$ = the expected number of accidents in each category.

If there is no discrepancy, λ^2 (chi-square) is equal to 0. As the discrepancy becomes larger, λ^2 becomes larger.

Table 2.3 is an abbreviated Chi-Square table in the cumulative form. From this table, it is possible to determine with a specified degree of certainty whether the sample results occurred purely by chance or whether they represent a statistically significant deviation from results which might be expected to occur by chance alone.

Table 2.3 λ^2 Factors for Testing Goodness of Fit

DEGREES OF FREEDOM DF	*90% CONFIDENCE LIMIT*	*95% CONFIDENCE LIMIT*
1	2.71	3.84
2	4.61	5.99
3	6.25	7.81
4	7.78	9.49
5	9.24	11.07
6	10.64	12.59
7	12.02	14.07
8	13.36	15.51
9	14.68	16.92
10	15.99	18.31
11	17.28	19.68
12	18.55	21.0
13	19.81	22.4
14	21.1	23.7
15	22.3	25.0
16	23.5	26.3
17	24.8	27.6
18	26.0	28.9
19	27.2	30.1
20	28.4	31.4

The format for applying the Goodness of Fit test is as follows:

1. State the ***null hypothesis,*** which in this case would be that the sample distribution is consistent with the distribution of results occurring purely by chance.

2. Establish a ***confidence limit,*** which defines the level of significance of the chi-square value which is considered to be a discrepancy between the sample and predicted distribution. A confidence limit of 90% or 95% is normally used.

3. Calculate λ^2 from Equation 2-6.

4. Compare the calcualted λ^2 with the appropriate value from Table 2.3. In using this table, the ***degrees of freedom,*** which represents the number of independent variables in the sample, is normally taken to be the actual number of sample variables minus one, or "n-1." If the calculated λ^2 value is greater than the appropriate value in Tablc 2.3, the null hypothesis is rejected, and the sample results are considered to be significantly discrepant from those expected to occur purely by chance. If, on the other hand, the calculated λ^2 value is equal to or less than the table value, the sample results are not considered to represent a significant variance from results predicted to occur by chance alone, and the null hypothesis is not to be rejected.

An example of the above procedure can be found in Simulation 2-1.

Another method for expressing the true mean from a small sample, for which a significant deviation between the sample mean and true mean would be expected, is to define bounds or a range within which the true mean would be expected to fall with a specified degree of confidence. This range is known as the ***confidence interval.*** Precalculated values derived from the Student *t* Distribution may be used to calculate the confidence interval of the true mean for the sample. Table 2.4 is adapted from the Student *t* Distribution, and may be easily used to establish the confidence interval for a given sample size and confidence limit of 90% or 95%. The limits of the interval are defined by the following:

SIMULATION 2-1

A biorhythm analysis of 3253 pilot-involved aircraft accidents resulted in the data summarized below. Did the accidents occur by pure chance, or is there a correlation with the pilot biorhythm cycles? Use a 90% confidence limit.

Days	*Number of Accidents**
Non-critical days	2557
Single critical physical *(P)*	240
Single critical sensitivity *(S)*	208
Single critical intellectual *(I)*	195
Double critical *(P)* and *(S)*	15
Double critical *(P)* and *(I)*	23
Double critical *(S)* and *(I)*	14
Triple critical *(P)*, *(S)* and *(I)*	1

**Source of Data:* Wolcott, "Biorhythms–Are They a Waste of Time?" *TAC Attack*, November 1975.

ANALYSIS PROCEDURE

1. *Null Hypothesis:* The accidents occurring on pilot biorhythm critical days could have occurred on those days purely by chance.
2. *Confidence Limit:* 90% (stated above).
3. *Calculation of Chi-Square:*

Day	*No. Accidents*	*Expected No. Accidents*†	$n_i - P_i n$	$\frac{(n_i - P_i n)^2}{P_i n}$
Non-Critical	2557	2590.8	-33.8	0.44
Critical *(P)*	240	246.7	-6.7	0.18
Critical *(S)*	208	199.3	8.7	0.38
Critical *(I)*	195	167.1	27.9	4.66
Critical *(P & S)*	15	19.0	-4.0	.84
Critical *(P & I)*	23	15.9	7.1	3.17
Critical *(S & I)*	14	12.9	1.1	0.09
Critical *(P, S & I)*	1	1.2	-0.2	0.03
Total (n)	3253			$\lambda^2 = 9.79$

† From Table 2.2, Probability (P_i) x 3253.

4. *Comparison with Table 2.3 Value:* Since 8 variables are being considered, *degrees of freedom* = 8-1 = 7. Referring to Table 2.3, with 7 degrees of freedom and a confidence limit of 90%, the appropriate value is 12.02.

CONCLUSION

Since the calculated 9.79 < 12.02, the null hypothesis cannot be rejected. Therefore, a significant correlation between pilot biorhythm criticality and pilot-induced accident occurrence was not demonstrated from the above data.

$$X + (a)s \geq \mu \geq X - (a)s \tag{2-7}$$

where: X = sample mean
s = sample standard deviation
a = value from Table 2.4
μ = true mean

In working with biorhythm data, another method for establishing whether a significant relationship can be shown between accident occurrence and biorhythm critical days based on a sample, is to determine the confidence interval for the true mean proportion of accidents occurring on critical days ("hits") to the total number of accidents. Then by expressing the proportions defining the interval as percentages, it may be determined from Table 2.2 whether the percentage of accidents which would be expected to occur by pure chance on critical days would fall within this range. The standard deviation, which is needed to establish the confidence interval, may be estimated by the following relationship:

$$s = \sqrt{\frac{A(1-A)}{n}} \tag{2-8}$$

where: s = standard deviation
A = ratio of "hits" to total number of accidents
n = sample size

The range, or confidence interval within which the true mean proportion of hits to total accidents would be expected to fall would then be defined by:

$$A + a\sqrt{\frac{A(1-A)}{n}} \geq \mu \geq A - a\sqrt{\frac{A(1-A)}{n}} \tag{2-9}$$

where: A = ratio of "hits" to total number of accidents
n = sample size
a = value from Table 2.4
μ = true mean proportion of hits to total accidents

Table 2.4 Factor "a" for Calculating % Confidence Limits [$\bar{X} \pm (a)s$]

SAMPLE SIZE N	*90% CONFIDENCE LIMIT "a" VALUE*	*95% CONFIDENCE LIMIT "a" VALUE*
2	4.50	9.00
3	1.70	2.50
4	1.15	1.59
5	.93	1.23
6	.81	1.04
7	.73	.93
8	.63	.83
9	.62	.76
10	.59	.72
11	.55	.69
12	.52	.63
13	.49	.60
14	.48	.58
15	.45	.55
16	.44	.53
17	.42	.51
18	.40	.49
19	.39	.48
20	.38	.47
21	.37	.45
22	.36	.44
23	.35	.42
24	.34	.42
25	.34	.41
26	.33	.40
27	.32	.39
28	.32	.38
29	.31	.37
30	.30	.36
40	.26	.31
60	.21	.25
120	.15	.18
N Greater than 120	$\frac{1.645}{\sqrt{N}}$	$\frac{1.96}{\sqrt{N}}$

Based on Student "t" Distribution.

Then, by converting proportions to percentages and referring to Table 2.2, it may be determined whether the percentage of hits predicted to occur on critical days by pure chance falls within the confidence interval of the true mean percentage of hits occurring on critical days in the sample. If the table value falls outside this confidence interval, it may be concluded that a statistically significant relationship between accident occurrence and biorhythm criticality has been demonstrated. Conversely, if the table value lies within the confidence interval, a relationship cannot be established based on the sample data.

An example problem which employs this methodology is given in Simulation 2-2.

SIMULATION 2-2

In a sample of 11 accidents, two occurred on biorhythm critical days. With a 95% confidence interval, do these results demonstrate a relationship between accident occurrence and biorhythm criticality?

ANALYSIS PROCEDURE

1. *Null Hypothesis:* The accidents recorded on biorhythm critical days could have fallen on those days purely by chance.
2. *Confidence Interval:* 95% (stated in problem).
3. *Calculation:* Utilizing equation (2-9):

$$A + \sqrt{\left(\frac{A(1-A)}{n}\right)}\ a \geq \mu \geq A - \sqrt{\left(\frac{A(1-A)}{n}\right)}\ a$$

determine the 95% confidence interval of the true mean proportion of hits to total accidents for the sample.

Given:

$$A = \frac{\text{total hits}}{\text{total accidents}} = \frac{2}{11} = .1818$$

n = sample size = 11
a = .69 (from Table 2.4)

$$.1818 + \sqrt{\left(\frac{.1818(.8182)}{11}\right)}\ .69 \geq \mu \geq .1818 - \sqrt{\left(\frac{.1818(.8182)}{11}\right)}\ .69$$

Simplifying,

$$.26 > \mu > .10 \quad \text{or} \quad 26\% > \mu > 10\%$$

4. *Comparison with Table 2.2 Value:* From the table, the percentage of accidents which may be expected to occur on critical days purely by chance is 20.4%. This value lies within the confidence interval defined above.

CONCLUSION

Since the table value lies within the sample 95% confidence interval, a relationship between accident occurrence and biorhythm critical days has not been shown, and the null hypothesis may not be rejected.

3

Biorhythm Analysis of 1300 Accidents

There have been numerous reports dealing with correlation of human performance with biorhythms, many of which have been unsubstantiated and have reported results ranging from no correlation to greater than 90% correlation. In contrast to the many poorly documented studies, the scientifically-based biorhythm analysis performed by Mr. Jacob Sanhein of the Naval Weapons Support Center, Crane, Indiana is based on more than 1,300 accidents and is supported by more than 500 pages of data. His report, "Biorhythm Analysis as Applicable to Safety" was presented to the National Safety Congress and Exposition on October 1, 1975. With Mr. Sanhein's permission, this report is presented in this chapter. To provide continuity, the table and figure numbers originally used have been changed to be consistent with the format of this text.

BIORHYTHM ANALYSIS AS APPLICABLE TO SAFETY

By Jacob M. Sanhein

ABSTRACT

Biorhythms, as referred to in this study, are the theoretical 12 day physical, the 28 day sensitivity and the 33 day intellectual, sinusoidal cycles which theoretically originate on the day a person is born and cycle uninterrupted until that person dies. Accident prone days or periods, are said to exist for approximately 24 hours, when any one of the three sinusoidal curves pass through zero degrees or one hundred and eighty degrees of the sinusoidal cycles. These accident prone periods represent approximately 20% of a person's life. Biorhythms are not intended to predict when an accident will occur, they only indicate a period of time when a person is four to seven times more prone to causing an accident. There is absolutely no fatalism or predestination involved with biorhythms.

This study reveals the results of evaluating 1308 accidents from three separate sources of accidents. The first source was industrial type accidents at NAVWPNSUPPCEN Crane, Indiana, the second source was community type accidents which were reported to the Bedford Medical Center Emergency Room, at Bedford, Indiana during the month of May 1975 and the third source was the single source Motor Vehicle Fatalities in the State of Indiana as reflected on the Official State Police records for the first seven months of 1974.

The primary objective of this study was to evaluate the number of "hits" and express this number as a percentage. A "hit" was scored when an accident occurred on the victim's biorhythm accident prone day or period. Therefore, this report documents exactly how the percentages of hits were derived. No attempt has been made to present any one side of biorhythms.

This study revealed that of the combined three sources of accident data, 41.75% of the accidents occurred on accident prone days and 52.03% occurred in accident prone periods.

A copy of the curves involved in this biorhythm evaluation and other supporting documents were forwarded to the National Safety Council for their archives. . . . The author of this report welcomes review of this data by other researchers of biorhythms.

I. INTRODUCTION

This is a report of the findings of a study which was conducted to test the theory of biorhythms. The theory being tested was that a large majority of accidents occur during periods of time which are predictable through the use of biorhythm curves. These theoretical predictable periods of time, in which a person is said to be several times more prone to causing an accident, are called "critical days." Because of the connotation associated with the word "critical" this report employs the term "accident prone day." Since accident prone days occur approximately 20% of the time (as many as seven days a month), it was arbitrarily established that the hit rate of 40% of the accidents occurring on accident prone days would indicate a successful amount of accident predictability through biorhythm evaluation. In practical safety terms this would mean by examining 20% of work time 40% of accidents could be eliminated.

This report refrains from explaining the basic concepts, calculations and approaches in the use of biorhythms. It is the intent of this report to explain how the various levels of hits were arrived at.

II. OBJECTIVE

There are two objectives of this report:

1. The primary objective of this report is to establish, define and publish the exact methods used to determine the percent of hits between accident occurrence dates and biorhythm accident prone days (also known as critical day).
2. The secondary objective of this report is to make suggestions for the unification of accident data used in biorhythm evaluations as well as to suggest revised terms, so that all evaluators can use the same values and language.

III. CONCLUSION

The results of the investigations, evaluations and testing involved in this study imply that approximately 42% of the accidents occur on accident prone days. This study was established to check for an approximate 40% rating of "hits," the results were an absolute minimum of 41.75% and the following conclusions were reached:

1. It is concluded that accidents do coincide with accident prone days 41.75% of the time.
2. It is concluded that uniformity in use of data is a requirement of biorhythm analysis.
3. It is concluded that the term "critical day" is not conducive in fighting the fear syndrome and that the term "accident prone day" should be used.
4. It is concluded that all evaluators should define their methods of evaluating "Hit Percentages" when reporting biorhythm analysis.

IV. RECOMMENDATIONS

The following recommendations resulted from this study:

1. It is recommended that additional evaluations of the biorhythm theory be conducted using a specially designed accident form, which will assist in obtaining more precise data and thereby provide a more refined evaluation.
2. It is recommended that all future safety biorhythm evaluations use uniform data which will comply with all four of the following requirements:
   a. That it was an accident.
   b. That it was a single source accident.
   c. That the victims' birth date is accurate.
   d. That the accident date is correct.

3. It is recommended that in future biorhythm evaluations the term "critical day" be dropped from use and the term "accident prone day" be used. The term "critical day" has a finality sound to it which frightens and misleads many people.
4. It is recommended that future evaluation reports contain information regarding the time duration of accident prone days or accident prone periods used in the evaluations.

V. SOURCE OF DATA

The information used in this report was obtained from three sources and the total number of accidents submitted for evaluation was 1308.

1. Eight hundred and twenty (820) industrial cases which reported to the base dispensary, at the Naval Weapons Support Center, Crane, Indiana during the year of 1974.
2. Two hundred and eighty-nine (289) community type accidents which occurred during the month of May 1975 and which were recorded at the 24 hour a day, Doctor manned, emergency room of the Bedford Medical Center, Bedford, Indiana.
3. One hundred and ninety-nine (199) fatal-single source-motor vehicle accidents as recorded by the Indiana State Police during the first seven months of 1974.

VI. TERMINOLOGY

In order to understand the reporting technique used in this report, two basic terms must be understood. These terms are the ***accident prone day*** and the ***accident prone period.*** It must be made clear that the sinusoidal biorhythm curves are considered to cross the abscissa at exactly midnight and the accident prone day exists for the next 24 hours. However, the 23 day and 33 day cycles cross the abscissa on half days at 180° of their cycle. The half day cross-over point is 12:00 noon.

1. *Method "A" Accident Prone Day.*
   In Figure 3.1 the "X" indicates the point when an accident occurred. In Method "A" the point at which the "X" is located and the point when the biorhythm curve crosses the abscissa, must be the same in order to be considered the "accident prone day." When the 23 day physical and 33 day intellectual curves change from active (+) to resting (-), then the accident prone day "X" is considered to be both Point 3 and 4 as shown in Figure 3.1, and defined as follows:
   a. Points 1, 3, and 6 marked X and Point 4 marked X_1 are accident prone days.
   b. Points 2, 5, and 7 marked X_2. are not considered accident prone days, in Method "A."

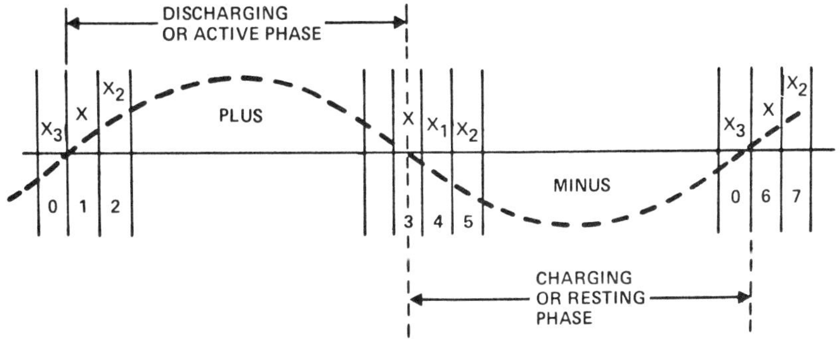

Figure 3.1 Accident Prone Days and Accident Prone Periods.

2. *Method "B" Accident Prone Period.*
   Accident Prone Periods (APP) differ from Accident Prone Days (APD) in that APP allows for time of birth, time of accident and for the fact that the 23 day physical and the 33 day intellectual cycles always

change from the active plus (+) stage to the resting minus (-) stage on a half day. This is indicated in Figure 3.1, Point 3. The approach that the APP allows for the time of birth and the time of accident widens the accident prone day into an accident prone period for the following reasons:

a. If person "A" was born at 12:01 a.m., 4 July 1960 his biorhythms started at 12:01 a.m., the exact moment that that person took his first breath. If person "B" was born at 11:01 p.m. on 4 July 1960 the biorhythms are theoretically 23 hours behind person "A". Both people, A and B, have approximately 24 hours accident prone periods, but person "B" lags person "A" by approximately 345°. This is indicated in Figure 3.2 in that the broken line indicates the curve for person "A" who was born at 12:01 a.m. The solid line indicates the curve for person "B". The section "A" of Figure 3.2 is displayed in Figure 3.3 in an enlarged view for easier explanation.

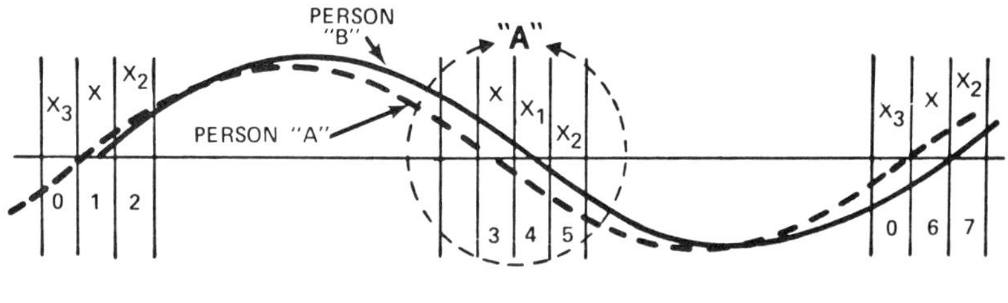

Figure 3.2 Theoretical 23 Hour Lag Allowing for Hour of Birth

b. It is shown in Figure 3.3 that for the 23 day cycle or 33 day cycle for person "A" (born 12:01 a.m.) the biorhythm accident prone 24 hour period X is from Point 3½ to Point 4½.

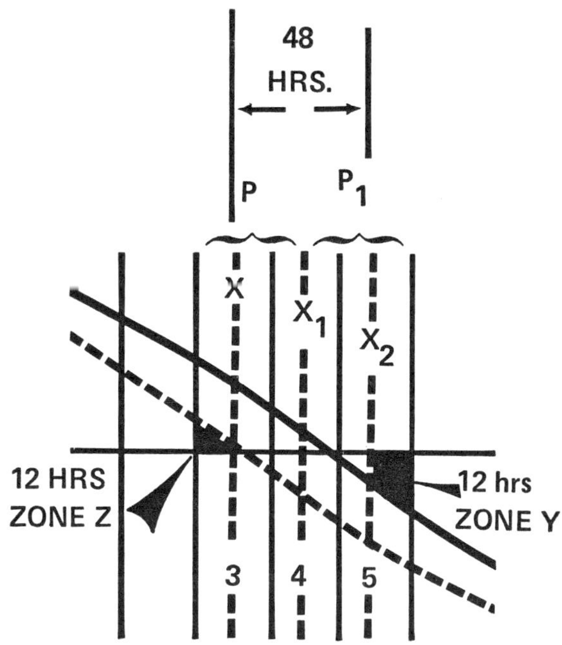

Figure 3.3 Half Day Accident Prone Day and Accident Prone Period

The solid curve shows that the person "B" (born 11:01 p.m.) has an accident prone period X_1 from almost Point 4½ to almost Point 5½. In the majority of accidents only the date of birth and date of accident is available. It is then difficult to pinpoint if the accidents (X, X_1, X_2) which occurs at Point 3, 4 and 5, is actually Person "A's" or person "B's" 24 hour period (P and P_1) because time of birth is not usually known. If exact time of accident is known, then time zone "Z" (12:00 midnight to 12:00 noon) Point 3 can be eliminated from analysis, as can Zone "Y" (12:00 noon to 12:00 midnight) of Point 5, for both person "A" and

"B". However this allows an approximate 48 hour period for both persons A and B because time of birth is not known. This period of doubt is cause for the accident prone period.

c. This report uses absolute time of birth and absolute time of accident to explain the theoretical differences between accident prone day and accident prone period. This should not be interpreted to imply that such accuracy can in reality be obtained. The human time clock cannot function at such a refined accuracy. The inaccuracy of the human function is very important in the truc evaluation of biorhythms for industrial accident prevention.

d. This analysis holds true for Points 1 and 2 of all three biorhythm curves and therefore creates a possible 48 hour APP. When 48 hour periods are used for the APP, in effect we are examining possibly 14 days of a month. This results from the regular approximate seven accident prone days in a month being doubled. The accident prone period used in this report involves consideration for the time the accident occurred and extends the regular 24 hour accident prone day into a 36 hour accident prone period.

3. *Method "C".*

This method allows for accidents which occur two hours before the accident prone period theoretically starts. Therefore the accident X_3 which is indicated at Point "0" of Figures 3.1 and 3.2 would be considered as occurring in the accident prone period if accident X_3 took place on or after 22 hours of Point "0".

VII. DATA CLASSIFICATION

The data which was used in this report compared three different types of data.

1. *Controlled Data* are those accidents from NAVWPN-SUPPCEN, Crane. It is said to be controlled because most accidents are limited to industrial accidents, involving only eight hours a day, five days a week with the largest percentage being males who are over 18 and less than 60 years of age.
2. *Pure Data* is considered to be those accidents from the Bedford Medical Center, because the accidents were reported to the emergency ward, 24 hours a day, seven days a week from all types of accidents without limitations of age, sex or occupation. Victims of accidents recorded in the month of May, 1975, at the Bedford Medical Center range from less than 2 months old to 80 years old.
3. *Limited Pure Data* is the motor vehicle data on fatalities from the State of Indiana, during the first seven months of 1974. This data was selected because the 28 day sensitivity cycle of the biorhythm theory is the least disputed cycle. This sensitivity cycle is said to affect your motion ability, a person's driving ability. This data was originally planned to be called "limited" because it was of the general public exposure type, with accidents occurring 24 hours a day, seven days a week. It was thought to be limited to an age bracket from 16 to 70 years of age and limited to single source accidents, involving only one vehicle and only the driver of the vehicle would be analyzed. The age range for these accidents was from 12 years old to 94 years old.
4. *Uniformed Data.* In order to use data in any biorhythm evaluation, that data must meet four basic

conditions before it can be counted as legitimate and be subjected to biorhythm analysis. The four conditions data must meet are:

a. That it was an accident.
b. That it was a single source accident.
c. That the victim's birth date is accurate.
d. That the accident data is accurate.

If any one of the above conditions cannot be authenticated then the data cannot be used in biorhythm analysis. The use of improper data can cause a reduction in the percentage of hits. In order to obtain 776 accidents for biorhythm evaluation, 1308 accidents were subjected to the four conditions of acceptable data and 532 cases could not be used.

5. *Examples of Uniformed Data.* The accident number 5-3-304 of the NAVWPNSUPPCEN Crane data is an example of not meeting condition "a". Accident number 5-3-304 showed up on the dispensary accident list as occurring on 3 May 1974. A check showed that this "accident" reported to the dispensary on 5 May 1974 with a reaction to a tetanus shot. The shot was given on 3 May 1974. This was not an accident and is not subject to biorhythm accident evaluation.

The satisfaction of condition "b" that it is a single source accident is ultimately important and must be considered in detail. Accident number 5-2-290 of NAVWPNSUPPCEN Crane data is an accident involving a laceration of the head of the victim. The laceration occurred on 2 May 1974, was reported on 2 May 1974, and it was an accident but is not usable in biorhythm analysis. The reason why it is not usable is because the laceration occurred when a wrench slipped out of the hand of a co-worker and hit the victim in the head. Since biorhythms have nothing to do with predestinations or fatalism, accidents which were caused by other than the victim cannot be used, unless the information on the person who caused the accident is authenticated. In such cases the biorhythm plot and evaluation would not be of the victim.

The exact birth data and date of accident is many times in error. Accidents reported several days late should be checked to authenticate exact date because people do guess at accident dates when they are reported late.

VIII. FINDINGS

Tables 3.1 and 3.2 indicate the following findings:

1. *NWSC 1974 Accidents.*
   Of the 820 industrial cases which reported to the NWSC Dispensary during 1974, 156 did not have birth day information and could not be plotted. This resulted in a balance of 664 accidents to be plotted and evaluated. Of the 664 cases, 217 did not meet the conditions of "Unified Accident Data" and were not counted in accident analysis. The remaining 447 accident cases of the original 820 reports to the base dispensary were plotted and analyzed and revealed the following results which are indicated in Table 3.1, Methods Results Comparison.
   a. 38.48% of the 447 accident cases occurred on non-accident prone periods using Method B.
   b. 61.52% of the 447 accident cases occurred on accident prone periods using Method B.
2. *Bedford Medical Center–May 1975 Accidents.*
   Of the 210 community type accidents for the month of May 1975 which were recorded in the emergency room of the Bedford Medical Center, 56 cases did not meet the conditions of "Unified Accident Data" and

Table 3.1 Methods Results Comparison

SOURCE OF DATA	*NO. OF TESTS*	METHOD A		METHOD B		METHOD C	
		NO. OF HITS	*%*	*NO. OF HITS*	*%*	*NO. OF HITS*	*%*
NWSC	447	189	42.37	275	61.52		
B.M.C.	154	60	38.96	92	59.74	111	72.07
M.V.F.	165	75	45.45	75	45.95	87	52.72
TOTAL	776	324	41.75	442	56.95	198	62.37

Table 3.2 Effect of Combining Methods "A" and "C"

	NO. OF TESTS	METHOD "A"		METHOD "C"		METHOD "A" & "C"	
		NO. OF HITS	*%*	*NO. OF HITS*	*%*	*NO. OF HITS*	*%*
BMC	154	60	38.96	19	12.33	79	51.30
MVF	165	75	45.45	12	7.27	76	52.72
TOTALS	319	135	42.32	31	9.71	166	52.03

were not subjected to analysis. The remaining 154 accidents were plotted and analyzed using Method "A" plus Method "C" and revealed the following results which are also shown in Table 3.2, Effect of Combining Methods "A" and "C".

a. 48.70% of the 154 accidents reported to the Bedford Medical Center Emergency Room, occurred on non-accident prone periods.

b. 51.30% of the 154 accidents reported to the Bedford Medical Center Emergency Room, occurred on accident prone periods.

3. *Fatal 1974 Motor Vehicle Accidents.*
Of the 199 fatal automobile accidents which occurred from 1 January 1974 through 30 July 1974, 34 did not meet the conditions of "Unified Accident Data" and were not used. Evaluation of the remaining 165 accidents are shown in Table 3.2 and revealed that:
   a. 47.28% occurred in non-accident prone periods.
   b. 52.72% occurred in accident prone periods.

IX. DISCUSSION

Over the past several years I have read and studied many articles on biorhythms. I have articles which indicate huge success with biorhythms and I have other articles which show complete failure and one article which provides a sensational sex life through the use of biorhythms. In all of these articles I have never read a detailed classification of terms, usage of date and comparison of different data. Therefore, this study was conducted to document exactly how the percentages of hits were derived. No attempt was made to present any one side of biorhythms. However, if biorhythms can be shown by other researchers, using the standards of this study, to have nearly the same percentages of hits then the findings of this study can be confirmed. Additional verification, using set standards of data could add to validation of the biorhythm theory. If this theory can be proven and practical proof through application of this theory be demonstrated, then for a 20% time examination, a large reduction in accidents could result.

The fact that biorhythms have only been evaluated against reduction of personal injury accidents is amazing. I propose a new form to investigate accidents by workers which create scrap, cause lost time and affect the quality of the product. A 40% to 50% reduction of scrap or damaged machinery, boats or equipment could sizeably affect the profit structure of a company. The approach of using biorhythm to affect a profit is indeed a much more involved study and should not be attempted by the novice. As a matter of fact, an evaluation of accidents using the biorhythm theory should not be attempted without detailed experience in the subject. The articles which I have read which treat biorhythms as a game are injurious to the theory and to the people who are mislead.

In the evaluation of the data contained in Table 3.1, Methods Results Comparison, the time factor must be considered. Method "A" is the pure biorhythm accident prone day which allotted approximately 20% of the time for affecting a large percentage of the accidents. Method "A" in Table 3.1 records a total "Hit Relationship" for the three data sources combined as 41.75%. Method B, in this table indicates a total 56.95% hit relationship which is an increase over Method "A" by an average 15.2%. Method "C" adds only two hours to either Method "A" or "B". In combining Method "C" to Method "A" the increase in time covered increases by 1.9% and the number of "Hit Relationships" increases by 9.71%. This is indicated in Table 3.2, Effects of Combining Methods A and C. This indicated that the period of a few hours before the APD has a direct relationship to the APP. This information also implies that the straightforward application of the biorhythm accident prone (critical) day does not exist. There are other factors such as group effect, personal preservance and environmental conditions.

It is desired that this report and study will be useful. The study did indicate some amount of biorhythm predictability but not to the extent of some of the reports that I have read. There is also some type of predominance existing or associated with the negative phase of the biorhythm cycles.

X. CREDITS

This report would not have been possible without the cooperation of the following individuals and organizations: Captain R. L. McCarthy, Commanding Officer, and Mr. Ralph Chipman, Operations Research Group Leader of the Data Processing Department, Naval Weapons Support Center, Crane,

Indiana; Superintendent Robert DeBard, Indiana State Police, Indianapolis, Indiana; Lt. John E. Ferguson, Accidents Records Section, Indiana State Police; Dr. Douglas Neil, Naval Postgraduate School, Monterey, California; and Mrs. Bonnie Clar, RN, Director of Nurses, Bedford Medical Center, Bedford, Indiana.

* * * * * *

OTHER STUDIES BY SANHEIN

In addition to the above study, Mr. Sanhein has performed several other recent investigations related to biorhythms which are of interest.

BIORHYTHMS AND SUICIDES

In a recent study Mr. Sanhein investigated 226 suicide cases in order to determine whether a correlation could be shown with biorhythm critical days. The suicides occurred over a ten year period in a city with a population over 150,000. The results of the experiment are summarized in Table 3.3.

Table 3.3 Summary of Results of Suicide Study by Sanhein

CRITICAL BIORHYTHM CYCLE	*NUMBER OF HITS*	*PERCENT OF TOTAL HITS*
Physical	25	46.29
Sensitivity	17	31.48
Intellectual	12	22.22

The total number of hits represents 54/226, or a percentage rate of 23.89. These results, although based on a fairly small sample, suggest that there is not a significant correlation between biorhythms and suicides.

BIORHYTHMS AND NATURAL DEATHS

In another experiment Mr. Sanhein studied 88 natural deaths in order to discover whether a correlation with biorhythms could be demonstrated. The results of this investigation are summarized in Table 3.4.

Table 3.4 Summary of Results of Natural Death Study by Sanhein

CRITICAL BIORHYTHM CYCLE	*NUMBER OF HITS*	*PERCENT OF TOTAL HITS*
Physical	12	54.14
Sensitivity	2	9.52
Intellectual	7	33.33

Here the total number of hits represents 21/81, or a percentage rate of 23.86. Once again the sample size for this investigation was relatively small. However, as with the suicide experiment, no correlation of natural deaths with biorhythm critical days was demonstrated.

SENSITIVITY OF DATA

In a separate study of 39 industrial accidents, Mr. Sanhein demonstrated that depending on the data and the time period used for the accident prone days, it is possible to arrive at a variety of results. In the sample of 39 cases submitted, only 28 cases were actually used in the analysis. This was due to the fact that even though 39 cases were reported as accidents, some could not be considered to be accidents as a result of human performance. (Three cases involved skin rashes, while eight involved intoxication.) Thus, the 28 remaining cases were subjected to biorhythm analysis.

If the total 39 cases were used in the analysis, a hit rate of 35.89% would result. When 36 cases are included (eliminating the skin rash cases), a hit rate of 36.11% is obtained. The results based on the 28 cases included 12 direct hits, or 42.85% coinciding with the theoretical biorhythm accident prone 24 hour periods. Here, the 28 cases are the most meaningful in the biorhythm analysis.

If a researcher wanted to show a high correlation, he might consider an accident falling on a day before or after the accident prone day to be a hit. In this situation, 21 hits would occur out of 39 tests for a hit rate of 53.84%. Out of a sample of 28 cases, the correlation would be even higher: a hit rate of 54.14%. Neither of these apparent higher correlation rates may be considered meaningful, however, because the accident prone period has been extended from 20% to approximately 60% of the time, and the number of hits expected to occur by chance alone would be increased.

4

A Comprehensive Analysis of Aircraft Accidents

As pointed out in Chapter 1, various investigators have reached conflicting findings regarding biorhythms. The conclusions reached by Sanhein leave little doubt as to the possibility of a relationship between biorhythm critical days and accidents. At the other end of the spectrum, Major John Wolcott and his associates have found no correlation between aircraft accidents and biorhythms. In this chapter the study, "Correlation of Occurrence of Aircraft Accidents with Biorhythmic Criticality and Cycle Phase" will be presented. This study is reprinted with permission of the Advisory Group of Aerospace Research and Development of the North Atlantic Treaty Organization, and was originally published in CONFERENCE PROCEEDINGS NO. 190. The study was prepared by Major John Wolcott, U.S. Air Force and Lt. Col. Robert R. McMeekin, U.S. Army, both of the Aerospace Pathology Division of the Armed Forces Institute of Pathology, Mr. Robert E. Burgin, Air Safety Investigator for the National Transportation Safety Board, and Dr. Robert E. Yanowitch, M.D., Chief of the Accident Investigation Branch, Office of Aviation Medicine, Federal Aviation Agency. The study utilizes a statistical approach for determining the probability of the results occurring purely by chance. The approach differs from the one used by Sanhein both in the definition of "accident prone days" and in the designation of which types of accidents should be included in the investigation. While some introductory information concerning biorhythm theory included here repeats that previously presented, the study by Wolcott, McMeekin, Burgin and Yanowitch is presented in its entirety to enable the reader to understand the complete basis of the investigation. Table numbers and figure numbers have been changed to allow consistency with the format of this book.

CORRELATION OF OCCURRENCE OF AIRCRAFT ACCIDENTS WITH BIORHYTHMIC CRITICALITY AND CYCLE PHASE[*]

By

Major John H. Wolcott, Lt. Col. Robert R. McMeekin, Robert E. Burgin, and Robert E. Yanowitch, M.D.

SUMMARY

The occurrence of aircraft accidents on various biorhythmic phases of cycles was studied. Aircraft-accident data were obtained from the National Transportation Safety Board for general civil aviation and from the U.S. Army Agency for Aviation Safety for military accidents. The accidents were divided into two groups, pilot- and nonpilot-involved cases, using the causal factors given by the respective accident boards. No correlation could be found between the occurrence of aircraft accidents and either the critical period, the negative phase, or the peak days of the negative phase of the biorhythmic cycles. Data were evaluated by chi-square analysis when considering all three cycles or the physical and emotional cycles alone, and all were studied with a critical period of 24 or 48 hours' duration.

[*] The opinions or assertions contained in this study are the private views of its authors and are not to be construed as official or as reflecting the views of the Department of the Air Force, of the Army, of Defense, of the National Transportation Safety Board, or of the Federal Aviation Agency.

INTRODUCTION

The "Biorhythm Theory" was developed from contributions of Swoboda, Fliess, and Teltscher. It predicted that behavior in terms of physical, emotional, and intellectual performance was influenced by cyclic variations. The work of the above authors is summarized in Thommen's book.[1] Briefly, there are three cycles of 23-, 28-, and 33-day duration that govern physical, emotional, and intellectual performance, respectively. Each cycle is described by a sine curve composed of a high or positive phase, a low or negative phase, and nodal points where the curve crosses the abscissa. Each cycle starts on the first positive phase at the moment of birth. The positive phases correspond to periods of better performance, the lower phase to periods of poorer performance, and the crossover or nodal points are termed "critical periods." The critical periods represent times of poorest performance and greatest susceptibility to accidents. Critical periods are usually defined to have a duration of 24 hours. Occasionally, 48- or 72-hour critical periods are found in the literature. Figure 4.1 presents each curve independently and appropriately identifies the critical 24- or 48-hour periods as well as high, low, and peak days of the positive and negative phases as used in this study.

From this theory, accidents would be expected to occur more frequently on critical days, days when more than one cycle is critical, days when one cycle is critical and the others are in a negative phase, or when all cycles are negative. Some proponents of the theory believe that the two most important cycles are the physical and emotional cycles when considering causes of accidents. Finally, one would expect that accidents involving an element of human error would be most likely to correspond to biorhythmic criticality.

The most frequently used method for testing the theory is to test previous accident data for correlation with the critical and low periods. This is possible because knowing only the birth date of the individual involved enables accurate calcula-

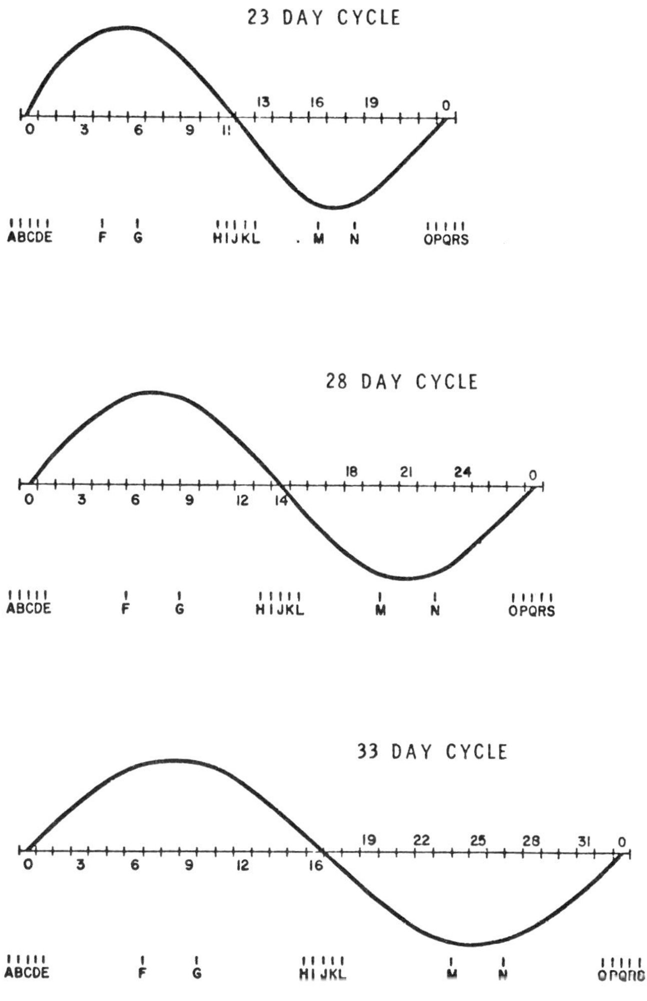

Physical (23 day), Emotional (28 day), and Intellectual (33 day) cycles. Letters represent the following in each graph: C, first cycle node or crossover point; B to D, 24-hour critical period; F to G, peak positive days: D to I, positive phase; J, second crossover point or half-cycle node; I to K, second 24-hour critical period; H to L, second 48-hour critical period; M to N, peak negative days; K to P, negative phase; Q, first crossover, new cycle; P to R, next 24-hour critical period; and O to S, next 48-hour critical period. (AFIP Neg. 75-9912.)

Figure 4.1 Biorhythm Cycle Definitions

tion of the state of each cycle on the date of the accident. The theory is most accurately tested when time of birth and accident are available. Frequently, time of birth is difficult to obtain and an average must be used.

Initial attempts to apply the theory to aircraft-accident data were summarized previously.[2] Data by Woodham, see Thommen,[1] Brady,[3] Weaver,[4] and Williamson[5] did indicate a possible role for biorhythms as a causal agent in aircraft accidents. Interpretation of these data was made difficult by small sample sizes, lack of clearly defined critical periods, and lack of appropriate statistical testing. Sacher[6] recently published a study of over 4,600 U.S. Navy flight accidents, all of which involved an element of pilot error and over 2,300 of which were termed "pilot-factor-only" by the accident board. He used appropriate statistical testing and a 24-hour critical period and found no significant deviation from random expected results. In a preliminary study[2] to this more extensive report, we also failed to find significant correlation between aircraft accidents and biorhythmic criticality or phase of cycle.

Two reports received recently have studied the correlation of biorhythms with industrial-type accidents. Mason[7] studied 13,285 lost-time accidents reported to the Workmen's Compensation Board of British Columbia during the 4-month period beginning 1 Jan. 1971. He used both the 11th and 12th days of the 23-day physical cycle and the 16th and 17th days of the 33-day intellectual cycle, which expanded his expected percentage from 20.37 to 26.59. He found that 3,584 accidents occurred on critical days, which is 26.98%. The study was carefully evaluated statistically, and no significant deviations from normal could be found. When a 5,259-member subgroup least likely to involve human error was studied, there were no significant deviations from the random biorhythm model or between the entire group and the smaller group.

Sanhein[8] studied 776 industrial and general accidents from 3 sources and found that 324 (41.7%) occurred on critical days. The critical period was apparently of 24 hours' duration. No breakdown as to type of critical days was given, and no statistical testing was reported.

MATERIALS AND METHODS

Source information was obtained from the files of the National Transportation Safety Board (NTSB)[9] and included the cause, date, and time of accident. Birth dates were obtained by manually searching individual accident records. U.S. Army data were obtained from computer files of the U.S. Army Agency for Aviation Safety, Fort Rucker, Alabama. Accidents were separated into groups based on accident-board determination of pilot involvement in accident causation. Accidents were listed as pilot involved when the NTSB data listed the first cause of accident as pilot factor. Accidents not so listed (mostly material failures) were considered as nonpilot involved. U.S. Army data listed pilot involvement as definite, primary or secondary role, suspected role, or not involved. The U.S. Army cases listed as suspected were not included in either of the other groups but were listed separately (199 cases) and included in the total number of cases.

The total days between dates on which the accident occurred and the birth date were calculated with either a Hewlett Packard HP 80 or HP 65 calculator. Total days between dates was then divided by 23, 28, and 33, using a simple program on the HP 65 calculator. An average time of birth of 12:00 noon was used. For accidents occurring outside the United States, the time of accident was converted from local time to Eastern Standard Time.

The critical period (day) was defined as a 24-hour period beginning 12 hours before the exact time that the curve crossed the abscissa. The first critical period in each cycle and the midpoint period of the 28-day cycle then started at 12:01 a.m. and ended at 12:00 midnight. The midpoint critical period for the

23-day physical and 33-day intellectual cycles occurs at 24:00 hours on the 11th or 16th day, respectively. The critical period began, then, at 12:01 p.m. the 11th or 16th day and ended 24 hours later.

The hypothesis was formed that biorhythm is not involved as a causal factor in aircraft accidents. Chi-square analysis[10] was used to test the difference in observed and expected values. Chi-square tables were constructed with an over-all probability of 1.0, except Table 4.7. Each individual chi-square value as well as the total value was studied for significance. Individual values exceeding 2.71 or 3.84 were significant at the p=0.1 and the p=0.05 level, respectively. A probability of 0.1 was used throughout this report and was deemed adequate for initial screening of accident data, although a higher level of significance would be required to insure a relationship between biorhythms and accident occurrence.

Table 4.1 presents the expected percentage occurrence for each of the 27 possible combinations of positive, negative, or critical phases (+ - C) that could occur among three cycles. The ratios 10.5/23, 13/28, and 15.5/33 define a positive or negative phase, and the ratios 2/23, 2/28, and 2/33 define a critical day for the 23-day physical, the 28-day emotional, and the 33-day intellectual cycles, respectively. Data were evaluated by considering all three cycles, by considering the physical and emotional cycles only, and by defining the critical period as 24 hours or as 48 hours.

Correlation of phase of cycle to accident occurrence was done by studying the number of accidents that occurred on various combinations of positive and negative phases for all three cycles.

Finally, a correlation between accident occurrence and the peak days of the positive or negative phase in each cycle was investigated. The exact peak times occur at 5.75, 7.00, and 8.25 days for positive and 17.25, 21.00, and 24.75 days for negative phases in the physical, emotional, and intellectual

Table 4.1 Expected Number of Accidents Occurring on Critical and Noncritical Days in the Biorhythmic Cycles

Criticality and Cycle	Cycle Phases P–E–I†	Expected Percentage	PILOT INVOLVEMENT IN ACCIDENT CAUSATION AND SOURCE OF DATA*				
			INVOLVED			NONINVOLVED	TOTAL
			NTSB	USA	Both	(NTSB & USA)	(All Cases)
NONCRITICAL	+ + +, – + –, + – +, + – –, – + +, – – +, – + –, – – –	9.9555 each	323.85	102.14	426.00	136.09	581.90
	TOTAL	79.6443	2,590.83	817.15	3,407.98	1,088.74	4,655.21
CRITICAL SINGLE PHYSICAL	C + +, C + –, C – +, C – –	1.8963 each	61.69	19.46	81.14	25.92	110.84
	TOTAL	7.5850	246.74	77.82	324.56	103.69	443.34
SINGLE EMOTIONAL	+ C +, – C –, – C +, – C –	1.5316 each	49.82	15.71	65.54	20.94	89.52
	TOTAL	6.1265	199.30	62.86	262.15	83.75	358.09
SINGLE INTELLECTUAL	+ + C, – – C, – + C, – – C	1.2846 each	41.77	13.18	54.97	17.56	75.08
	TOTAL	5.1383	167.15	52.72	219.87	70.24	300.33
DOUBLE P & E	C C +, C C –	0.2917 ea	9.49	2.99	12.48	3.99	17.05
	TOTAL	0.5835	18.98	5.99	24.97	7.98	34.11
DOUBLE P & I	C + C, C – C	0.2447 ea	7.96	2.51	10.47	3.35	14.30
	TOTAL	0.4894	15.92	5.02	20.94	6.69	28.61
DOUBLE E & I	+ C C, – C C	0.1976 ea	6.43	2.03	8.46	2.70	11.55
	TOTAL	0.3953	12.86	4.06	16.91	5.40	23.11
TRIPLE	C C C	0.0376	1.22	0.39	1.61	0.51	2.20

**NTSB* = National Transportation Safety Board,[9] cases from 1972 general civilian accidents; *USA* = U.S. Army, cases from U.S. Army Agency for Aviation Safety during period 1969 to 1975.

†*P* = physical or 23-day cycle; *E* = emotional or 28-day cycle; *I* = intellectual or 33-day cycle; + = positive phase of cycle; – = negative phase of cycle; C = critical day.

cycles, respectively. Peak days were chosen to include either a 48- or 72-hour period surrounding the peak times. Peak positive days were as follows: days 5 and 6, physical cycle; days 6, 7, and 8, emotional cycle; and days 7, 8, and 9, intellectual cycle. Peak negative days were as follows: days 17 and 18, physical cycle; days 20, 21, and 22, emotional cycle; and days 24, 25, and 26, intellectual cycle.

RESULTS

CORRELATION WITH BIORHYTHMIC CRITICAL DAYS

If biorhythm is involved as a causal factor in aircraft accidents, the most logical finding should be that a larger than expected number of accidents occur on critical, especially multiple critical days of the physical, emotional, and intellectual cycles. The number of accidents occurring on each possible combination of cycle phases is presented in Table 4.2.

Table 4.3 presents data derived from both NTSB and U.S. Army cases subdivided into categories of pilot involved, noninvolved, and total. The first section of the table separates accidents as occurring on noncritical days, each combination of singly or doubly critical days, and triply critical days. The 12.82 total chi-square value for the smallest sample size (U.S. Army pilot-involved category) is the only total value that is significant. The second section of the table combined the individual singly and each type of doubly critical days into two groups and presents a total chi-square value based on 3 degrees of freedom. None of the total chi-square values is near the value of 6.25 required for significance.

Although not listed as a separate table, the total chi-square values derived by using each of the 27 possible combinations of phase of cycle shown in Tables 4.1 and 4.2 also result in 5 total chi-square values well below the significant level for a table with 26 degrees of freedom. Seven individual chi-square values of the

Table 4.2 Observed Numbers of Aircraft Accidents Occurring on Each Combination of Cycle Phases

SEQUENCE NUMBER	CYCLE PHASES P–E–I†	PILOT INVOLVEMENT IN ACCIDENT CAUSATION AND SOURCE OF DATA*					
		Involved			*Noninvolved*	*Suspected*	TOTAL
		NTSB	USA	BOTH	(NTSB & USA)	(USA)	(ALL CASES)
1	+ + +	342	96	438	126	21	585
2	+ + –	315	101	416	138	24	578
3	+ – +	328	106	434	134	24	592
4	+ – –	337	120	457	125	13	595
5	– + +	318	100	418	139	18	575
6	– – +	293	90	383	161	21	565
7	– + –	325	91	416	141	23	580
8	– – –	299	103	402	123	16	541
9	C + +	62	20	82	18	1	101
10	C + –	57	17	74	22	6	102
11	C – +	58	26	84	29	7	120
12	C – –	63	22	85	25	4	114
13	+ C +	63	17	80	22	1	103
14	+ C –	47	22	69	21	1	91
15	– C +	51	16	67	21	2	90
16	– C –	48	25	73	17	1	91
17	+ + C	48	11	59	18	3	80
18	+ – C	42	9	51	27	3	81
19	– + C	48	5	53	22	2	77
20	– – C	56	16	72	20	5	97
21	C C +	8	1	9	2	0	11
22	C C –	7	1	8	3	1	12
23	C + C	11	3	14	4	0	18
24	C – C	12	1	13	5	0	18
25	+ C C	8	2	10	3	1	14
26	– C C	6	4	10	1	0	11
27	C C C	1	1	2	0	1	3
TOTAL		3,253	1.026	4,279	1,367	199	5,845

**NTSB* = National Transportation Safety Board,[9] cases from 1972 general civilian accidents; *USA* = U.S. Army, cases from U.S. Army Agency for Aviation Safety during period 1969 to 1975.

†*P* = physical or 23-day cycle; *E* = emotional or 28-day cycle; *I* = intellectual or 33-day cycle; + = positive phase of cycle; – = negative phase of cycle; C = critical day.

Table 4.3 Chi-Square Values Derived from Comparison of Aircraft Accidents on Critical and Noncritical Days of the Physical, Emotional, and Intellectual Cycles

CRITICALITY AND CYCLE[†]	PILOT INVOLVEMENT IN ACCIDENT CAUSATION AND SOURCE OF DATA[*]				
	INVOLVED			NONINVOLVED	TOTAL
	NTSB	USA	BOTH	(NTSB & USA)	(ALL CASES)
NONCRITICAL (1-8)	0.44(–)[**]	0.13(–)	0.57(–)	0.00	0.42(–)
CRITICAL					
SINGLE					
PHYSICAL (9-12)	0.18(–)	0.66(+)	0.00	0.91(–)	0.09(–)
EMOTIONAL (13-16)	0.47(+)	4.67(+)	2.75(+)	0.09(–)	0.80(+)
INTELLECTUAL (17-20)	4.31(+)	2.61 (–)	1.04(+)	4.00(+)	4.00(+)
DOUBLE					
P–E (21-22)	0.84(–)	2.66(–)	2.54(–)	1.11(–)	3.62(–)
P–I (23-24)	3.15(+)	0.21(–)	1.75(+)	0.80(+)	1.91(+)
E–I (25-26)	0.10(+)	0.93(+)	0.56(+)	0.36(–)	0.15(+)
TRIPLE P–E–I (27)	0.04(–)	0.95(+)	0.09(+)	0.51(–)	0.29(+)
TOTAL CHI-SQUARE	9.53	12.82	9.30	7.78	11.28
TABULAR CHI-SQUARE FOR EACH OF ABOVE COLUMNS, p = 0.1, n = 7				12.02	
NONCRITICAL	0.44(–)	0.13(–)	0.57(–)	0.00	0.42(–)
CRITICAL					
ALL SINGLE P, E, and I	1.45(+)	0.82(+)	2.23(+)	0.07(+)	1.86(+)
ALL DOUBLE PE, PI, EI	0.38(+)	0.62(–)	0.02(+)	0.21(–)	0.04(–)
TRIPLE P–E–I	0.04(–)	0.95(+)	0.09(+)	0.51(–)	0.29(+)
TOTAL CHI-SQUARE	2.31	2.52	2.91	0.79	2.61
TABULAR CHI-SQUARE FOR EACH OF ABOVE COLUMNS, p = 0.1, n = 3				6.25	

[*] *NTSB* = National Transportation Safety Board,[9] cases from 1972 general civilian accidents; *USA* = U.S. Army, cases from U.S. Army Agency for Aviation Safety during period 1969 to 1975.

[†] *P* = physical or 23-day cycle;
E = emotional or 28-day cycle;
I = intellectual or 33-day cycle;
numbers in parentheses ( ) refer to left-hand column numbers of Table 4.2 and to the corresponding cycle phase in Table 4.1, where source material is found.

[**] (–) = refers to fewer than expected numbers of accidents contributing to chi-square;
(+) = refers to more than expected numbers of accidents contributing to chi-square

40 presented in the first section of Table 4.3 exceeded 2.71; none of the 20 in the second section of the table did. One of the seven resulted from fewer than expected numbers of accidents occurring on doubly critical days of the physical and emotional cycles. Using a 27-member chi-square table (data not shown, but can be calculated from data presented in Tables 4.1 and 4.2), we found that 13 of 120 individual values exceed 2.71, 3 of these because of fewer than the expected number of accidents occurred on a particular type of critical day or more than the expected number occurred on noncritical days.

Thommen[1] himself–and, according to him, Schwing–believed that the physical and emotional cycles were of primary importance in determining accident causation. All possible combinations of phases of these two cycles were investigated and the data presented in Table 4.4. In the first section none of the total chi-square values exceeded the level required for significance with 8 degrees of freedom. The only individual values exceeding the 2.71 level were 2.75 for fewer noncritical days (– –) than expected and 2.72 for more accidents occurring on (+C) days in the total pilot-involved group, 3.98 for more (-C) days from U.S. Army pilot-involved group, and 2.92 from the total cases for fewer than expected numbers of doubly critical days.

In the second section the various types of noncritical, physical critical, and emotional critical are combined. One total chi-square value was significant (U.S. Army, pilot-involved group). The individual values for emotional critical (U.S. Army, pilot involved), emotional critical (both, pilot involved) and the same doubly critical value was significant.

Expanding the critical period to 48 hours resulted in 8-member chi-square tables (data not presented but would have same construction as Table 4.3) with total chi-square values slightly less than 11 for both pilot- and nonpilot-involved groups. Both these total values, particularly the nonpilot-involved one, included individual chi-square values resulting from fewer than

Table 4.4 Chi-Square Values Derived from Comparison of Aircraft Accidents on Critical and Noncritical Days of the Physical and Emotional Cycles Only to Expected Numbers

	*PILOT INVOLVEMENT IN ACCIDENT CAUSATION AND SOURCE OF DATA**				
	INVOLVED			*NONINVOLVED*	*TOTAL*
CYCLE PHASE (+, –, OR C) 23, 28 DAY (SOURCE)†	*NTSB*	*USA*	*BOTH*	*(NTSB & USA)*	*(ALL CASES)*
+ + (1, 2, 17)	0.35(+)**	0.41(–)	0.04(+)	0.21(–)	0.01(+)
+ – (3, 4, 18)	0.44(+)	1.41(+)	1.35(+)	0.05(–)	0.68(+)
– + (5, 7, 19)	0.00	2.12(–)	0.44(–)	0.52(+)	0.04(–)
– – (6, 8, 20)	2.50(–)	0.33(–)	2.75(–)	0.70(+)	1.04(–)
C + (9, 10, 23)	0.01(–)	0.05(–)	0.04(–)	2.27(–)	0.95(–)
C – (11, 12, 24)	0.02(+)	1.39(+)	0.49(+)	0.26(+)	1.09(+)
+ C (13, 14, 25)	1.34(+)	1.70(+)	2.72(+)	0.05(+)	1.59(+)
– C (15, 16, 26)	0.01(–)	3.98(+)	0.79(+)	0.70(–)	0.01(+)
C C (21, 22, 27)	0.87(–)	1.78(–)	2.16(–)	1.43(–)	2.92(–)
TOTAL CHI-SQUARE	5.54	13.17	10.78	6.19	8.33
TABULAR CHI-SQUARE FOR EACH COLUMN ABOVE, p = 0.1, n = 8			13.36		
NONCRITICAL	0.02(–)	0.55(–)	0.23(–)	0.19(+)	0.02(–)
CRITICAL PHYSICAL	0.00	0.46(+)	0.12(+)	0.49(–)	0.00
CRITICAL EMOTIONAL	0.55(+)	5.45(+)	3.21(+)	0.19(–)	0.93(+)
DOUBLE CRITICAL P–E	0.87(–)	1.78(–)	2.16(–)	1.43(–)	2.92(–)
TOTAL CHI-SQUARE	1.44	8.24	5.72	2.30	3.87
TABULAR CHI-SQUARE FOR EACH COLUMN ABOVE, p = 0.1, n = 3			6.25		

**NTSB* = National Transportation Safety Board,[9] cases from 1972 general civilian accidents; *USA* = U.S. Army, cases from U.S. Army Agency for Aviation Safety during period 1969 to 1975.

†numbers in parentheses ( ) refer to left-hand column numbers of Table 4.2 and to the corresponding cycle phases in Table 4.1, where source material is found.

**(+) refers to more than expected numbers of accidents contributing to chi-square;
(–) refers to fewer than expected numbers of accidents contributing to chi-square.

expected numbers of accidents occurring on certain types of critical days. There were 502 of 36.72% of the accidents occurring on critical days in the nonpilot-involved group and 1,629 or 38.07% in the pilot-involved group. The expected percentage is 37.78, and neither deviated significantly from this expected value.

Table 4.5 summarizes the individual and total chi-square values derived when the contribution of each of the three cycles —physical, emotional, and intellectual—was broken down into its three component parts—accidents occurring on positive, negative, or critical days. The data for the physical cycle do not include any total or individual chi-square values of significance. Significant chi-squares include one total value (6.15 for U.S. Army pilot-involved data) and one individual value (3.37, same source) for more accidents occurring on emotionally critical days than expected. Three of the individual chi-square values were significant in that more accidents occurred on intellectually critical days than expected.

CORRELATION OF ACCIDENT OCCURRENCE WITH PHASE OF CYCLE

Data presented in Table 4.6 were derived by breaking the data in Table 4.2 down into four categories:

(i) all three cycles positive;
(ii) any two cycles positive, the other either negative or critical;
(iii) one positive, the others negative or critical; and
(iv) all cycles negative or critical.

This enables evaluation of the contribution of multiple phases of cycle on accident causation. Neither a total nor any individual chi-square value was significant. The total values for the combined pilot-involved group and for the total cases were extremely low, 0.82 and 0.33, respectively.

Table 4.5 Chi-Square Values Derived from a Comparison of Occurrence of Aircraft Accidents on Positive, Negative, or Critical Phases of the Physical, Emotional, or Intellectual Cycle to Expected Values

CYCLE, PHASE, AND SOURCE OF DATA†	PILOT INVOLVEMENT IN ACCIDENT CAUSATION AND SOURCE OF DATA*				
	INVOLVED			NONINVOLVED	TOTAL
	NTSB	USA	BOTH	(NTSB & USA)	(ALL CASES)
PHYSICAL					
+ (1-4, 9, 10, 13, 14, 25)	1.36(+)**	0.52(+)	1.88(+)	0.16(–)	0.96(+)
– (5-8, 11, 12, 15, 16, 26)	1.14(–)	0.72(–)	1.81(–)	0.70(+)	0.64(–)
C (9-12, 21-24, 27)	0.05(–)	0.09(+)	0.00	0.99(–)	0.17(–)
TOTAL CHI-SQUARE	2.55	1.33	3.69	1.85	1.77
EMOTIONAL					
+ (1,2,5,7,9,10,17,19,23)	0.16(+)	2.20(–)	0.14(–)	0.07(–)	0.12(–)
– (3,4,6,8,11,12,18,20,24)	0.33(–)	0.58(+)	0.02(–)	0.32(+)	0.03(+)
C (13-16, 21, 22, 25-27)	0.19(+)	3.37(+)	1.64(+)	0.60(–)	0.17(+)
TOTAL CHI-SQUARE	0.68	6.15	1.80	0.99	0.32
INTELLECTUAL					
+ (1,3,5,6,9,11,13,15,21)	0.02(–)	0.20(–)	0.11(–)	0.15(+)	0.00
– (2,4,7,8,10,12,14,16,22)	0.59(–)	0.84(+)	0.05(–)	1.14(–)	0.62(–)
C (17-20, 23-27)	6.16(+)	1.67(–)	2.35(+)	3.55(+)	5.66(+)
TOTAL CHI-SQUARE	6.77	2.71	2.51	4.84	6.28
TABULAR CHI-SQUARE FOR EACH COLUMN ABOVE, p = 0.1, n = 2			4.61		

*NTSB = National Transportation Safety Board,[9] cases from 1972 general civilian accidents; USA = U.S. Army, cases from U.S. Army Agency for Aviation Safety during period 1969 to 1975.

†numbers in parentheses ( ) refer to left-hand column numbers of Table 4.2 and to the corresponding cycle phases in Table 4.1, where source material is found.

**(+) refers to more than expected numbers of accidents contributing to chi-square;
(–) refers to fewer than expected numbers of accidents contributing to chi-square.

Table 4.6 Chi-Square Values Derived from Comparison of Occurrence of Aircraft Accidents on Various Combinations of Phases of Cycle to Expected Numbers

CYCLE, PHASE, AND SOURCE OF DATA†	*PILOT INVOLVEMENT IN ACCIDENT CAUSATION AND SOURCE OF DATA**				
	INVOLVED			*NONINVOLVED*	*TOTAL*
	NTSB	*USA*	*BOTH*	*(NTSB & USA)*	*(ALL CASES)*
ALL THREE POSITIVE (1)	1.02(+)**	0.37(–)	0.34(+)	0.75(–)	0.02(+)
ANY TWO POSITIVE (2,3,5,9,13,17)	0.07(+)	0.00	0.06(+)	0.03(–)	0.03(+)
ANY ONE POSITIVE (4,6,7,10,11,14,15,18,19, 21,23,25)	0.22(–)	0.18(–)	0.39(–)	1.75(+)	0.01(+)
ALL NEGATIVE/CRITICAL (8,12,16,20,22,24,26,27)	0.21(–)	1.34(+)	0.03(+)	1.39(–)	0.27(–)
TOTAL CHI-SQUARE	1.52	1.89	0.82	3.92	0.33
TABULAR CHI-SQUARE FOR EACH COLUMN ABOVE, p = 0.1, n = 3			6.25		

**NTSB* – National Transportation Safety Board,[9] cases from 1972 general civilian accidents; *USA* = U.S. Army, cases from U.S. Army Agency for Aviation Safety during period 1969 to 1975.

†numbers in parentheses ( ) refer to left-hand column numbers of Table 4.2 and to the corresponding cycle phases in Table 4.1, where source material is found.

**(+) refers to more than expected numbers of accidents contributing to chi-square;

(–) refers to fewer than expected numbers of accidents contributing to chi-square.

The chi-square values presented in Table 4.7 were derived by comparing the number of accidents occurring on either peak negative or peak positive days of the three cycles. Both the observed numbers and the chi-square values are presented. Expected numbers can be computed by multiplying the total number of cases from each group by 2/23 for peak physical, 3/28 for peak emotional, and 3/33 for peak intellectual days.

There were no significant individual chi-square values for either peak positive or peak negative days of the physical cycle. Fewer than expected numbers of accidents occurring on peak negative days of the intellectual cycle resulted in a significant chi-square value of 3.22. There were more than expected numbers of accidents occurring on both peak positive and peak negative days of the emotional cycle, and both chi-square values were significant. One total chi-square value was significant but probably has little meaning since no trend toward more accidents occurring on either peak positive or peak negative days was evident.

DISCUSSION

ACCIDENT OCCURRENCE AND BIORHYTHMIC CRITICALITY

If the biorhythm theory does play a part in accident occurrence, then statistical testing of observed occurrences should be highly significant. If one performs chi-square analysis of the data reported by Thommen[1] and conducted by Schwing and Bochow, one will find that the values are extremely high. From Schwing's data a total chi-square of 850 would be computed for the following accident breakdown: 299 noncritical, 322 singly, 74 doubly, and five triply critical. For Schwing's data on death, 197 of 300 occurred on some type of critical day, and a chi-square value of 180 would be computed. From a statement by Thommen that triply critical days occur once per year, it is

Table 4.7 Chi-Square Values Derived from Comparison of Occurrence of Aircraft Accidents on Peak Positive and Peak Negative Days of the Physical, Emotional, and Intellectual Cycles to Expected Numbers

CYCLE	PHASE AND PEAK DAYS	PILOT INVOLVEMENT IN ACCIDENT CAUSATION AND SOURCE OF DATA*									
		INVOLVED						NONINVOLVED		TOTAL	
		NTSB		USA		BOTH		(NTSB & USA)		(ALL CASES)	
		OBS.	X^2 †	OBS.	X^2	OBS.	X^2	OBS.	X^2	OBS.	X^2
PHYSICAL	+ DAYS 5,6	286	0.03(+)**	88	0.02(–)	374	0.01	128	0.70(+)	515	0.09(+)
	– DAYS 17,13	296	0.61(+)	80	0.95(–)	376	0.04(+)	104	1.36(–)	503	0.05(–)
EMOTIONAL	+ DAYS 6–8	342	0.12(–)	94	2.31(–)	436	1.10(–)	173	4.31(+)	628	0.00
	– DAYS 20–22	339	0.26(–)	94	2.31(–)	433	1.41(–)	170	3.78(+)	625	0.00
INTELLECTUAL	+ DAYS 7–9	288	0.20(–)	99	0.35(+)	387	0.01(–)	132	0.48(+)	537	0.06(+)
	– DAYS 24–26	271	2.07(–)	99	0.35(+)	370	0.93(–)	106	2.39(–)	490	3.22(–)
TOTAL CHI-SQUARE			3.29		6.29		3.50		14.32		3.42
TABULAR CHI-SQUARE FOR EACH COLUMN ABOVE,	p = 0.1, n = 1						2.71				
	p = 0.1, n = 5						9.24				

*NTSB = National Transportation Safety Board,[9] cases from 1972 general civilian accidents; USA = U.S. Army, cases from U.S. Army Agency for Aviation Safety during period 1969 to 1975.

†OBS. = observed number; X^2 = chi-square.

**(+) refers to more than expected numbers of accidents contributing to chi-square;
(–) refers to fewer than expected numbers of accidents contributing to chi-square.

suspected that Bochow used a 48-hour critical period. Even so, based on the percentage of singly, doubly, and triply critical days reported, the total chi-square value (48-hour critical period) exceeds 11,000.

In the data presented here, from an evaluation of the contribution of all three cycles to accident occurrences on critical days, no ***total*** chi-square values were significant at the 0.05 level. This includes doing a complete set of tables (27-member chi-square tables) as well as the combined tables presented in Table 4.3. The only total chi-square value significant at the 0.1 level was from the smallest sample size, the U.S. Army pilot-involved group. In this case the individual chi-square value for noncritical-day accidents was only 0.13, and a large share of the total chi-square resulted from fewer than expected numbers of accidents occurring on singly critical days of the intellectual and on doubly critical days of the physical and emotional cycles. In the data presented in the second section of Table 4.3, when all singly and doubly critical days were combined, the total chi-square values were much less than the 6.25 required for significance. None of the individual chi-square values were significant.

If the contribution of only the physical 23-day and emotional 28-day cycle is analyzed, there is one total chi-square value that is significant at either the p=0.1 or p=0.05 level. This value, 8.24, results primarily from more than expected numbers of accidents occurring on singly critical days of the emotional cycle. There were fewer than expected numbers of accidents occurring on doubly critical days in data from all sources that contributed to each total chi-square value. The chi-square values for accidents occurring on all noncritical days were very low for data from all sources.

It is obvious from the chi-square value of 0.00 for critical days in the physical cycle (Table 4.5) for the combined pilot-involved group that the criticality of this cycle was exactly as expected from the random model. The one individually significant value for accidents occurring on critical days of the emotional cycle occurred in the U.S. Army pilot-involved group (Table 4.5 is not very high for the combined pilot-involved group, and is almost negligible for the total group). Three of the chi-square values for accidents occurring on critical days of the intellectual cycle were significant, one in the NTSB pilot-involved group, one in the nonpilot-involved group, and one in the combined group. This is also demonstrated in Table 4.3, in which the individual chi-square value for the intellectual cycle was 4.00. In actual numbers it means that 15 more accidents occurred on singly critical days and 9 more on doubly critical days involving the intellectual cycle for the combined pilot-involved group. Of the total cases considered, 35 more accidents occurred on singly critical days and 9 more on doubly critical days of the intellectual cycle than expected.

The contribution of this one cycle was, as shown before, not significant enough to influence the total chi-square value. If we also consider the data from Sacher[6] for over 4,600 U.S. Navy cases, it is found that almost exactly the expected number of accidents occurred on intellectually critical days. A combined table presented previously,[2] shows a resultant individual chi-square value of 0.88. Our conclusion is that the deviation from random found in this cycle is not significant. It seems unreasonable to assume that this cycle, which in our data accounts for almost all the deviation from expected values, is a causal factor of accidents.

In analyzing data by chi-square at the 0.1 level, it is expected that 1 in 10 chi-square values could exceed the value required for significance without invalidating the hypothesis. Of the individual chi-square values, there were 220 presented in Tables 4.3-4.7, or 355 if the total 27-member chi-square table is included (data not published here). Of these, 18 of 220 and 31 of 355 had chi-square values that exceeded 2.71. Tables 4.3, 4.4 and 4.5 were primarily concerned with an evaluation of criticality and accident occurrence, and here the ratio was 15 individually-significant critical values of the 170 presented. Twelve of these values were derived because more accidents than expected occurred on critical days (or fewer than expected numbers on noncritical days). In summary, although there were a few more accidents occurring on various types of critical days, the over-all deviation is not significant.

CONTRIBUTION OF PHASE OF CYCLE

Analysis of the contributions of positive versus negative phase for each cycle (Table 4.5) revealed that there were no individual chi-squarc valucs of significance (for + or -, not critical). Data presented in Table 4.6, which summarizes the combined phase of cycle, also had very low individual and total chi-square values: only 0.82 for the combined pilot-involved group and 0.33 for the total cases. A triply negative day had no different effect than a triply positive day.

From the 27-member chi-square table (not presented but can be calculated from values presented in Tables 4.1 and 4.2, lines 9 and 12), it can be determined that 63 accidents occurred on days when the physical cycle was critical and the other two were negative. There were 62 accidents occurring when the physical cycle was critical and the other two positive, the expected value for each type being 61.7. There were 63 accidents occurring on + C + days and 48 on - C - days (Table 4.2, line 13 and line 16, respectively). There were 48 on + + C and 56 on - - C days (lines 17 and 20, respectively, of Table 4.2). Again, no particular influence of phase of cycle can be demonstrated.

If negative phase of cycle is important, it might also be expected that accidents would occur more frequently than expected on days at the lowest point of each cycle, or peak negatibe days. Table 4.7 presents this information. For the combined pilot-involved cases (4,279), no individual values were significant. In the noninvolved group, more accidents than expected occurred on peak positive and peak negative days of the emotional cycle. The only value significant when all cases were considered was the 3.22 value resulting from fewer than expected numbers of accidents occurring on peak negative days of the 33-day cycle. No evidence supporting peak negative days as a possible causal factor was demonstrated.

CONCLUSION

No significant correlation could be found between occurrence of aircraft accidents and biorhythmic critical periods or negative phases of the pilots. This was true when evaluating all three cycles or only the physical and emotional cycles and with either a 24- or 48-hour critical period. Although a few more accidents occurred on intellectual-critical days, when our data for pilot-involved cases are combined with Sacher's,[6] this individual value is not significant. The number of individually significant chi-square values is less than the 10% that could occur using chi-square at the p=0.1 level without invalidating the null hypothesis. The extremely low chi-square values derived when studying combined phases of cycle and the lack of correlation of accident occurrence to peak negative days strongly negates the hypothesis that the negative phase of cycle(s) is an important part of accident causation.

Although a safety-oriented program to continually brief pilots on biorhythms may, in fact, reduce accident rates, a similar program of frequent personnel briefings on safety without mention of biorhythms may have the same effect. Briefing biorhythms is ill founded. Pilots are not more likely to have accidents on specific biorhythmic days.

ACKNOWLEDGMENTS

Grateful appreciation is extended to the Aerospace Pathology staff for their ceaseless efforts in computing, summarizing, and rechecking these data. The staff includes MSgt Robert G.

Howell and TSgt Robert A. House, as well as the secretaries, Mrs. Kay Weaver and Mrs. Hilda Giampetroni. We are indebted to Dr. Walter Foster for assistance in statistical computations.

REFERENCES

1. Thommen, G., *Is This Your Day?* New York, Crown Publishers, Inc., 1973.
2. Wolcott, J. H., McMeekin, R. R., Burgin, R. E., and Yanowitch, R. E., "Biorhythms–Are They a Waste of Time?" *TAC Attack,* Vol. 15, No. 11, pp. 4-9, 1975.
3. Brady, T., "Bio-What?" *TAC Attack,* Vol. 12, No. 3, pp. 16-19, 1972.
4. Weaver, C. A., "The Question of Ups and Downs." *U.S. Army Aviation Dig.,* Vol. 20, No. 1, pp. 13-17, 1973.
5. Williamson, T. B., "Cancel Today, and Save Tomorrow," *Air Scoop,* Vol. 15, No. 3, pp. 18-20, 1975.
6. Sacher, D., "The Influence of Biorhythmic Criticality on Aircraft Mishaps," Master's thesis, Navy Postgraduate School, Monterey, California, 1974.
7. Mason, K. (Workmen's Compensation Board of British Columbia), "An Investigation of the Biorhythm Theory," 5255 Heather Street 13, B. C.
8. Sanhein, J. M. (Quality Assurance Department, Naval Weapons Support Center, Crane, Indiana), "Biorhythm Analysis as Applicable to Safety," presented at the National Safety Congress and Exposition, Oct. 1, 1975 (unpublished report).
9. National Transportation Safety Board, Aircraft Accident Reports, brief format, "U.S. Civil Aviation: 1972 Accidents," NTSB BA-73-2, 3, 4, 6, 7, and 10, Washington, D. C., NTSB.
10. Dixon, W. J. and Mason, F. J., Jr., *Introduction to Statistical Analysis,* New York, McGraw-Hill, Inc., 1957, pp. 226-227.

5

Behavioral Analysis and Biorhythms

The initial thrust of biorhythm research has been in the area of correlating historical accident data with the predicted accident prone days of the biorhythm cycles. Current research is determining if a correlation exists between the spacial relationships of the three cycles and performance. One such research project is detailed in Chapter 6, dealing with the influence of the positive phase of the intellectual cycle and its effect on academic performance.

If relationships can be proven, the applications in this new area seem almost limitless. Industrial psychology, management and individual development are just a few possible applications.

BIO-ANALYSIS

The behavioral tendencies which may be associated with each of the three biorhythm cycles are summarized in Table 5.1. As previously pointed out, these descriptors do not predetermine behavior, but rather indicate a tendency which an individual may exhibit.

The overall biorhythmic behavior of an individual is a combination of the influence of each of the three cycles. The combination of positive, negative, and zero cycles for the physical, sensitivity and intellectual cycles result in twenty-four behavior-

Table 5.1 Summary of Biorhythm Behavior

CYCLE	DAYS	DESCRIPTION
PHYSICAL	2 – 11 1, 12,* 13* 14 – 23	PEAK DAYS ACCIDENT PRONE DAYS EBB DAYS
SENSITIVITY	2 – 14 1, 15 16 – 28	HARMONIOUS DAYS ACCIDENT PRONE DAYS STRESS DAYS
INTELLECTUAL	2 – 16 1, 17,* 18* 19 – 33	CREATIVE DAYS ACCIDENT PRONE DAYS BELOW PAR DAYS

*Accident Prone Days on half crossings start at 12 noon and end 12 noon the following day. All other Accident Prone Days start at 12 midnight and end at 12 midnight the following day.

al influence possibilities. For days other than critical days, or 80% of the time, the positive and negative portions of the cycles result in eight combinations of the physical, sensitivity and intellectual, as illustrated in Table 5.2.

BIORHYTHM MANAGEMENT

A good manager overcomes working loss or error in order to maximize profits. The manager is a planner and strives to achieve numerous objectives. "Management by exception" is a term which expresses poor management. It means that the manager has executed his plan to achieve a goal, except that this or that unplanned event happened. Exceptions in the performance of work means inadequate planning, inadequate relationships between management and the worker, or suboptimal performance of the worker causing him to be more prone to work errors.

The essence of biorhythm management is establishing a set of procedures by which the biorhythmic accident prone periods

Table 5.2 Combinations of Biorhythm Behavior*

STATE OF CYCLE			*DESCRIPTION OF BEHAVIOR*
P	*S*	*I*	
+	+	+	The individual should be at a peak period physically and be more apt to be creative and in a good mood.
–	+	+	The individual should be in a creative cycle and in a good mood, but be in an ebb in the physical cycle.
+	–	+	The individual should be in a peak physical cycle and be in a creative intellectual phase, but also be experiencing a stressful period.
–	–	+	The individual is in a creative cycle but is at an ebb physically and emotionally.
+	+	–	The individual is in a peak physically and in a harmonious mood while below par intellectually.
–	+	–	The individual is in an ebb period both physically and intellectually while in a positive emotional state.
+	–	–	The individual is below par intellectually, feels stressful but is at a peak in the physical cycle.
–	–	–	The individual is below par intellectually and physically and is in a stressful mood.

*NOTE:
P = Physical Cycle
S = Sensitivity Cycle
I = Intellectual Cycle
\+ = Cycle in Positive Phase
– = Cycle in Negative Phase

of each individual worker may be considered and compensated for on an individual basis. Through this procedure the likelihood of worker-performance types of accidents may be minimized.

BIORHYTHM COMPATIBILITY

People's desire to be compatible is evidenced by commercial services such as computer dating services, growth in astrology and unproven biorhythm compatibility predictions. Computer biorhythm programs are already used to predict the degree of compatibility with one's mate.

Some biorhythm theorists claim that individuals with similar biorhythms are highly compatible while others with different cycles have a low compatibility. There appears to be little substance to this theory. In fact, it is not clear whether compatibility would be indicated by matching or opposite biorhythm cycles. For example, business partners whose intellectual cycles rise and fall together might make a lot of bad decisions during mutual low periods. On the other hand, management theory indicates that a certain degree of difference or conflict is a necessary ingredient for brainstorming and creativity.

If biorhythms are proven to become a valuable tool in the area of compatibility, it is likely that the value would be in understanding a colleague's or mate's behavior. It would be worthwhile to understand actions of associates due to irritability, fatigue, and moodiness in terms of predictable biological rhythms.

ACCIDENTS DO NOT JUST OCCUR

An accident may be defined as an unplanned, noncontrolled and undesirable event or sudden mishap which interrupts an activity or function. Not all accidents result in an injury. In fact, only a few of all accidents do. However, even if an injury is not involved, the result of the accident could cause a delay in production, damage of material, reduction of quality or additional wear on equipment. All told, virtually all accidents result in some cost to the company.

There are many contributing causes to accidents as illustrated in Figure 5.1. Any accident can be thought of as four distinct parts:

- contributing causes
- immediate causes
- the accident
- the results of the accident

The ***contributing factors*** are comprised of the accumulation of unsafe acts and unsafe conditions which provided circumstances predisposed to accident occurrence. The supervisor plays an important role in preventing accidents. Effective safety management necessitates recognizing and controlling these contributing factors.

The safety responsibilities of the supervisor lie in several specific areas:

- Recognizing unsafe acts of employees
- Recognizing poor performance of employees
- Enforcing safety rules
- Planning safety as part of the job
- Correcting hazards due to improper maintenance, etc.
- Providing clear safety instructions
- Insuring that safety devices are worn

Thus, ineffective supervision is itself a potential contributing factor to an accident.

The worker is the second contributing factor in an accident. The mental, physical, and emotional conditions of the person result in the following:

Mental
Lack of safety awareness
Improper attitude
Inattention

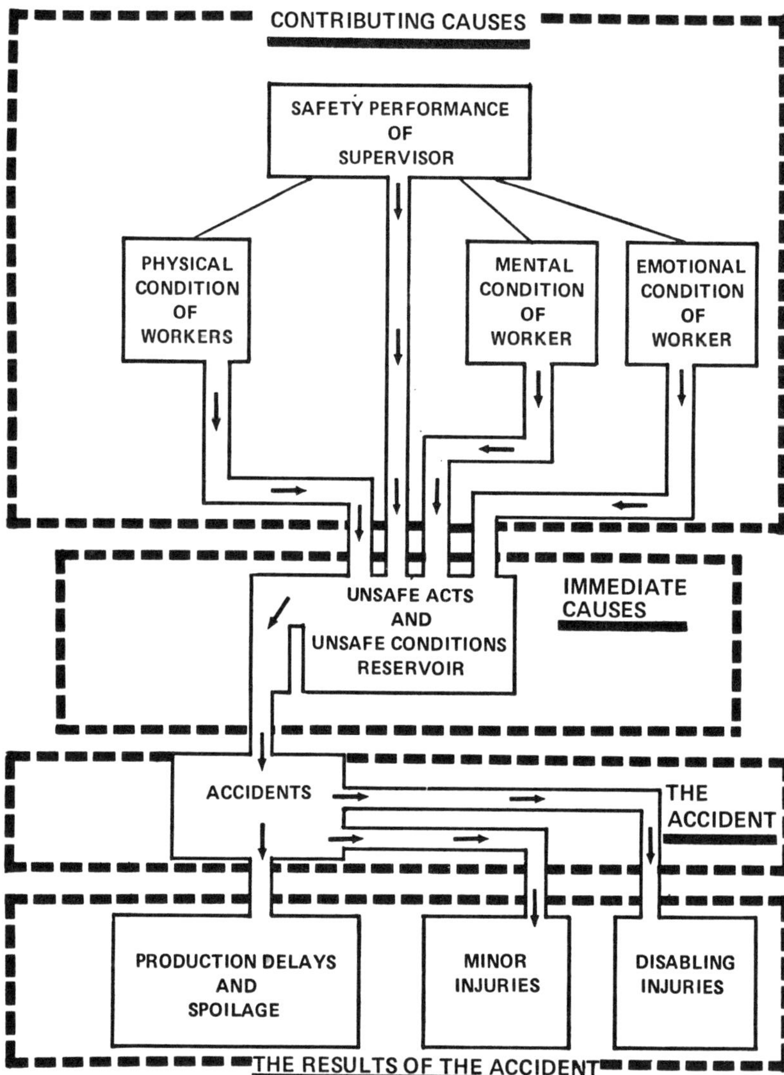

Figure 5.1 Anatomy of an Accident Flow Chart

Slow mental reaction
Lack of coordination

Physical
Extreme fatigue
Lack of coordination
Physically unqualified for job

Emotional
Lack of emotional stability
Nervous
Temperamental

The ***immediate causes*** of the accidents include unsafe acts such as not using protective equipment or proper tools, and hazardous movements. Accidents are also caused by unsafe conditions such as defective equipment, improper illumination, ventilation, hazardous housekeeping and ineffective safety devices.

Today's safety programs emphasize elimination of unsafe conditions. Yet it was recognized by Heinrich[1] as far back as 1931, that 88 percent of all accidents result primarily from unsafe acts, 10 percent from unsafe conditions, and 2 percent from "acts of God," or unpreventable circumstances.

When an accident does occur it indicates that something has gone wrong. The supervisor who wants to prevent accidents needs to have a working knowledge of the fundamental steps that can be taken to control the contributing causes of the accident. One of the key steps recognized to prevent accidents is for the supervisor to develop a safety awareness. The traditional approach aims at conditioning a person's mind so that whenever there is an element of danger present, a sixth "safety sense" will help the person react in a safe way. Safety awareness is developed by training repetition, and constant correction. The added dimension to safety awareness is that "off days" in human behavior do not just occur on a random basis. The biorhythm model can be used to determine lows in human performance and accident prone days. The supervisor can then reinforce "safety consciousness" at specified times to individuals who are more likely to be a contributing cause in an accident.

There have been classic studies which confirm that psychological factors influence safety. Dr. Schulzinger[2] and other researchers indicate that

- Twice as many injuries occur to young males as opposed to young females.
- Youth is a factor which increases the likelihood of accidents.
- Poor temperament, attitude and physical condition of a worker increase the likelihood of an accident.

The concept of "accident proneness" has long been recognized in safety programs. A small percentage of all workers are indeed more accident prone or have a higher incidence of injuries. By definition "accident proneness" is described as a combination of human attributes which make a person highly inclined to bring about accidents, regardless of the environment, at a rate higher than statistically expected by chance alone.

"Accident prone days" as applied to the biorhythm theory have a similar meaning. The key difference is that the individual can compensate for an accident prone day by being aware of the state of the body.

Thus the supervisor who emphasizes the physical, mental and emotional states of the workers has a better chance of reducing in-plant accidents. Present research with biorhythms and worker performance indicates that the "model" is not perfect.

There are numerous factors which are considered to affect employee performance, including:

- Morale
- Attitude
- Work environment
- Stress

- Fatigue
- Bad habits

Identifying all of the primary factors which influence performance is a complex task, and it is evident that no single factor alone can be a perfect predictor of a worker's or a manager's performance or susceptibility toward making errors. Biorhythms provide a potential management tool to aid in better decision making by considering the periods when a worker may be most susceptible to errors.

BIORHYTHM COMPUTER ANALYSIS

With the current public interest in biorhythms, over one hundred firms throughout the United States offer services directed toward calcualting biorhythms for individuals. These services range from machines in cocktail lounges which will plot one's daily biorhythm for twenty-five cents to computer services advertised in newspapers, magazines and on television which will compute monthly or annual biorhythm charts. Often these biorhythm analyses tend toward sensationalism and predict behavior well beyond areas which have been verified or investigated by biorhythm researchers.

Three biorhythm charts, Figures 5.2, 5.3 and 5.4, are presented on the following pages with typical "bio-analysis" such as may be provided by these commercial services.

BIO-ANALYSIS NO. 1 (Figure 5.2)

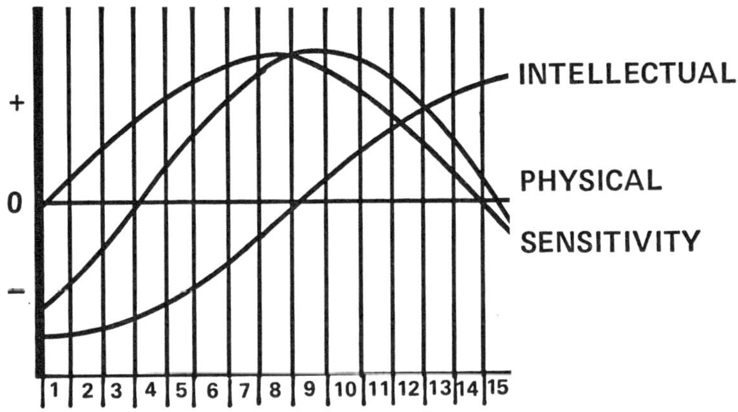

BEGINNING PHASE

Plus Period – Physical and Sensitivity cycles.
Minus Period – Intellectual cycle.

DAY 1 The individual is sluggish mentally and should be careful of his actions especially those involving his impulses.

DAYS 2 & 3 The individual will experience a more positive feeling and be in a better mood. Recall is still not at its best.

DAYS 4 & 5 This accident prone period is coupled with poor reflexes and a slow thinking response.

DAYS 6 THROUGH 14 The individual should excell in areas which require endurance, a good attitude. Some mental sluggishness may still be experienced in the first half of this period. Decisions should be weighed carefully during Days 9 and 10, but mental reflexes should be sharp during days 11 through 14.

DAY 15 This accident prone period is coupled with a day when both the physical and sensitivity cycles change. Extra caution should be taken to avoid accidents and the intellectual power should be used to avoid any hazards.

BIO-ANALYSIS NO. 2 (Figure 5.3)

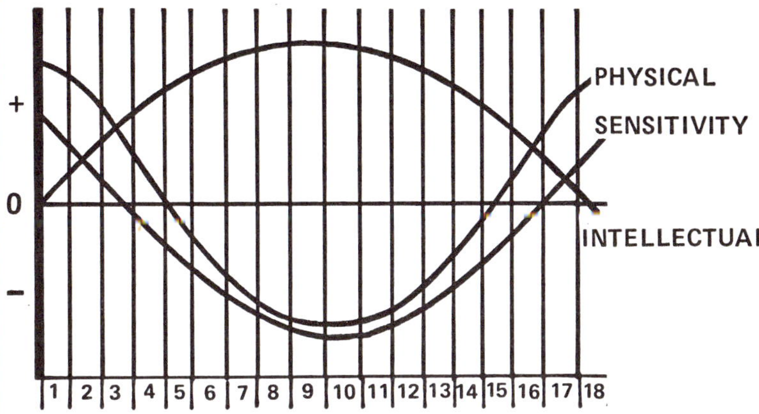

BEGINNING PHASE

Plus Period – Intellectual cycle.
Minus Period – Sensitivity and Physical cycle.

DAYS 1 & 2 The individual should weigh decisions carefully Day 1 since the intellectual cycle changes from low to high. Since both the physical and sensitivity cycles are in the positive period the overall state of the individual should remain good.

DAYS 3 & 4 This accident prone period is coupled with the individual's power of quick recall. The individual should make a conscious effort to think before he acts.

DAYS 5 THROUGH 14 This period finds the individual less tense and less eager to get things done. It is a period of low physical and sensitivity rhythms. The individual's tendency to give in easily will be counterbalanced by the high intellectual phase.

DAYS 15 & 16 This accident prone period is associated with characteristics that the individual will have a tendency to be sluggish while the sensitivity cycle is low. The individual's sharp mind can overcome sluggish reactions.

DAY 17 This accident prone period is coupled with both the intellectual and sensitivity periods changing phase while the physical cycle is full of energy. The individual should be careful not to allow the physical high lead to impulsive action, especially since good judgement may be lacking during this period.

BIO-ANALYSIS NO. 3 (Figure 5.4)

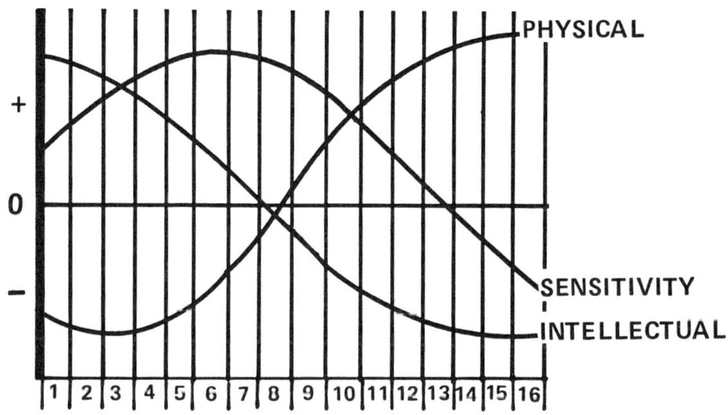

BEGINNING PHASE

Plus Period – Intellectual and Sensitivity cycle.
Minus Period – Physical cycle.

DAYS 1 THROUGH 7 During this period the individual's intuition and creative ability are highly stimulated. This is a period to avoid boring physical chores.

DAYS 8 & 9 This accident prone period is a result of a switch from high to low in the individual's intellectual rhythm followed by the physical cycle changing phase.

DAYS 10 THROUGH 13 These are days when the individual will be at his best physically and emotionally. The individual will need to think carefully since the intellectual period is at a low. This period is characteristic of self-confidence.

DAY 14 This accident prone day is coupled with reflexes which cannot match the strength of the individual. Since the intellectual cycle is at a low, this would be a good period to concentrate on routine activities and non-demanding tasks.

REFERENCES

1. Heinrich, W. H., *Industrial Accident Prevention,* McGraw-Hill, 1931.
2. Schulzinger, S. M., *The Accident Syndrome,* Charles C. Thomas Co., 1958.
3. Schwartz, G. and Robert, P. E., "A Look at the Matter of Susceptibility to Work Errors as Related to Biorhythm," *Professional Safety,* April 1976.
4. Cohen, Daniel, *Biorhythms in Your Life,* Fawcett Publications, 1976.
5. DeReamor, R., *Modern Safety Practices,* John Wiley & Sons, Inc., 1958.

6

Statistical Techniques for the Analysis of Biorhythms

Chapter 5 suggests that if an objective methodology can te used, it would be possible to accurately demonstrate whether or not a relationship exists between the biorhythm cycles and human performance. One such attempt at a quantitative analysis of the biorhythm cycles as they relate to human variability was presented in the paper "Statistical Analysis of the Theory of Biorhythms" by Douglas E. Neil, Louis J. Giannotti and Thomas A. Wyatt of the Man-Machine Systems Design Laboratory of the Naval Postgraduate School, Monterey, California. Appreciation is expressed to the authors for permitting the presentation of their report, which constitutes the remainder of this chapter. As in Chapters 3 and 4, table and figure numbers have been modified, and minor format changes have been made to permit consistency within this text.

STATISTICAL ANALYSIS OF THE THEORY OF "BIORHYTHMS"

By Douglas E. Neil, Louis J. Giannotti and Thomas A. Wyatt

SUMMARY

This study describes the development of a statistical technique for the analysis of biorhythms. The model developed is applied in the analysis of academic performance and industrial accidents. The results suggested a possible relationship between the hypothesized intellectual cycle and academic performance, and between the physical cycle and industrial accidents. The findings indicated that "good" and "poor" academic performance were distribured differently, in that the distribution of "below average" academic performance was more complex than either "average" or "above average" performance. The analysis of the criticality phenomena also suggested a possible relationship between physical critical days and accident occurrence.

INTRODUCTION

Research into human variability has evolved as a result of the scientific interest of researchers and the implications the understanding of the topic holds for a technical society. The ability to predict human efficiency in the operation and maintenance of complex and potentially hazardous man-machine systems and in decision making situations has economic as well as humanitarian considerations.

Biological periodicity represents one aspect of biological functioning which is pertinent to gaining an understanding of the variable character of human behavior. Colquhoun[1] has suggested that most, if not all, biological functions are cyclic in nature. Considerable research has been devoted to (a) identifying specific biological cycles, and (b) relating biological periodicity to physiological and/or psychological/performance variation (Alluisi and Chiles,[2] Wever,[3] Hauty,[4] Hauty and Adams,[5, 6] Lobban,[7] Lobban and Tredre,[8] etc.). Generally, the investigations have concentrated on cycles for which objective physiological manifestations exist (e.g. menstrual cycle), or cycles which appear to originate with entrainment to specific environmental influences (e.g. circadian periodicity). Oatley and Goodwin[9] have interpreted the results of such investigations as suggesting that biological periodicity represents a "strategy" whereby the living organism attempts to maintain consistency and stability in the dynamic organization of the living process. Therefore, in cycles which appear to function independently of the environment, Oatley and Goodwin have postulated that the rhythmicity represents a means whereby two incompatible functions can be accomplished (e.g. cell growth and division). Cycles which are, at least partially, environmentally determined are considered as adaptive responses on the part of the organism to the surrounding environment. In either case, biological periodicity can be considered as integral to the living process with physiological and/or psychological/behavior manifestations.

One theory of biological periodicity for which little scientific interest has been generated is the theory of biorhythms. A contributing factor to the lack of interest and experimentation by the scientific community is the basis upon which the theory has been formulated. As discussed by Thommen,[10] the theory essentially originated with the subjective interpretations of observations of human behavior. Unlike the better established biological oscillations, the cycles cannot be attributed to entrainment to environmental factors, nor have physiological concomitants to the theoretical curves been identified.

The theory suggests that man's behavior is characterized by three inherent cycles. The cycles include a 23 day physical cycle, and 28 day emotional or sensitivity cycle, and a 33 day intellectual cycle.[10] As the names imply, the physical cycle influences tasks of a physical nature, the emotional curve takes on increased importance in situations of high emotional content, and the intellectual cycle is of particular importance in pursuits requiring cognitive activity. According to the theory, the cycles begin at birth and terminate with death. Each cycle begins at zero at the moment of birth and progresses upward until a maximum positive value is reached. Maximum positive value requires 5.75 days for the physical, 7 days for the emotional or sensitivity cycle, and 8.25 days for the intellectual. Once the maximum value is achieved, the cycles begin the return to the zero point, reaching zero in a time-span equivalent to the time required to reach maximum. The cycles cross the base line or zero point and continue into the negative portion of the cycle. The time-frames are the same as for the positive portion of the oscillation.

Theoretically, the cycles continue in a relatively fixed manner throughout life. When a cycle is in the positive portion of oscillation, performance potential is enhanced on tasks requiring the capacities related to the particular rhythm. For example, if the physical cycle is in the positive period, performance potential on physical types of tasks is increased. During periods in which a cycle is in the negative phase, performance potential is considered degraded and the probability of poor performance increased. On days during which a cycle is crossing from positive to negative, or from negative to positive, performance potential is considered highly unstable. These crossover days have been characterized as "critical days," and Thommen[10] has suggested they are "full of danger or difficulty." (See Figure 6.1.)

The theory has been applied to a variety of situations including athletic contests (Wallenstein and Roberts,[11] Thommen[10]), historical events (Thommen[10]), and accidents, illness and safety (Willis[12]). Generally, the approach used has involved the examination of isolated events in relation to specific segments of a particular cycle or cycles. The results have been interpreted as suggesting a degree of predictive association between the event(s) of interest and the theoretical biorhythm.

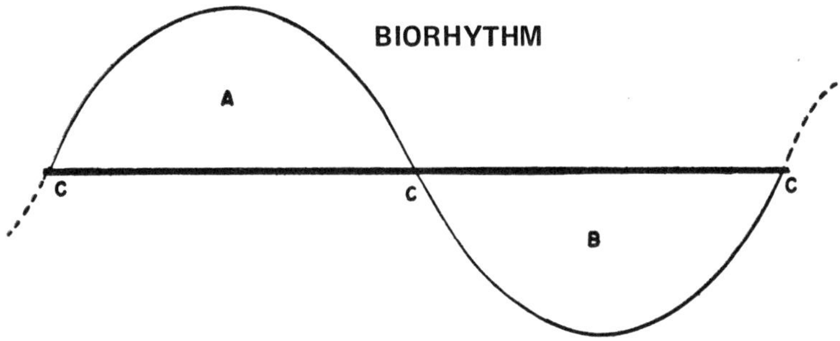

A = Days on which there is a predisposition for above average performance.

B = Days on which there is a predisposition for below average performance.

C = Critical Day.

Figure 6.1 Biorhythm Human Variability

However, as a result of the subjectivity involved in the interpretation of the events, and the tendency to limit the investigations to a specific segment of the hypothesized cycles, the results have not provided sufficient information upon which to evaluate the theory or the concept it purportedly represents.

The primary objective of this study was to develop a statistical model whereby an attempt to statistically validate the theory could be accomplished. The approach used here involved attempts to (a) define the distributional form of observed performance data, (b) determine the presence or absence of a biorhythmic influence on observed performance, and (c) measure the extent of the relationship between observed performance and the theoretical biorhythm.

METHOD

INTERPRETATION OF THE THEORY

As a result of numerous ambiguities in the theory, it was considered essential that a precise statement of the investigators' interpretation of the theory represent the first step in the investigation. The oscillations associated with the theory were considered as biological strategies whereby the cyclically organized biological system can provide the energy required for biological functioning. Therefore, the rhythmicity was considered as an essential aspect of the process of energy expenditure and conservation or restoration. At another conceptual level, the oscillations can be thought of in terms of a polarization process. As such, the periods specified by the theory can be considered as variations in the polarity of the physical, emotional and intellectual subsystems. Thus, at a given time a particular subsystem may be responsive in one way to the environment, while during another time-frame, the response may be quite different given the same stimulus. Therefore, the notion of biorhythms may imply an alteration of sensory thresholds during different time-spans. If this is the case, the feedback loops may be affected, and performance of the total system may be impaired or enhanced, depending upon the match of the organism-environment whole.

As such, the theory of biorhythms was considered in a framework similar to other, well-established biological periodicities which are essential to the dynamic living process. Biorhythms were not considered to be predictors of the future nor as the sole or even primary variable in explaining the present. The hypothesized cycles were viewed as one of the many variables which contribute to the day-to-day variation in human functioning. Biorhythms were considered to represent a predisposing factor which contributes to limiting an individual's capacity in much the same way fatigue, motivation or circadian rhythms might influence an individual's potential capacity. Further, in order for the potential to be realized or tested, the situation must sufficiently challenge that capacity, otherwise the influence could be masked through extra effort on the part of the performer.

According to the theory, critical days were considered as periods of instability. The interpretation of biorhythms used in this study would suggest that it is on these days that the organism's polarity is in a state of flux and therefore, the feedback process is highly variable. Thus, the organism is not receiving immediate, accurate information regarding its organismic state. The variability attributed to this time-frame can be interpreted as a response to the varying sensory thresholds and an inability on the part of the individual to accurately assess his capacity.

DEVELOPMENT OF THE STATISTICAL MODEL

Once the investigators' interpretation of the theory had been specified, the most appropriate statistical techniques for the analysis had to be selected. Various parametric and nonparametric procedures were considered. In view of the objectives of this study, procedures which would provide an analysis of specific segments of the hypothesized periodicities were considered inappropriate in that they were capable of offering, at best, a partial analysis of the total theory.

Therefore, it was felt that an appropriate statistical model should include a measure of the relationship or dependency of observed "good" and "poor" performance days, and biorhythmically predictable "good" and "poor" performance days. If such days are not equally likely, the distribution should be other than uniform. If the latter is observed (i.e. non-uniform distribution) a technique for identifying the observed distributional form and comparing it to the theoretical biorhythms should be included.

The model selected included the nonparametric techniques of (1) Chi-Square-Contingency-Test, (2) Chi-Square-Coefficient of Contingency, and (3) Chi-Square-Goodness-of-Fit-Test as the primary statistical procedures (Siegel[13]).

CHI-SQUARE-CONTINGENCY-TEST

A two-way Chi-Square-Contingency-Test was selected to investigate the relationship between the day of occurrence of an observed event and the sequential day of a given biorhythm. If, as the theory suggests, an individual's capacity to perform is enhanced during the positive portion of a cycle and degraded during the second or negative portion, a relationship may exist between observed performance and the sequential day of a cycle. The Chi-Square-Contingency-Test was selected to analyze the possible relationship between observed performance and biorhythms.

CHI-SQUARE-COEFFICIENT-OF-CONTINGENCY

Assuming a relationship or apparent dependency is observed between performance and biorhythms, the Chi-Square-Coefficient-of-Contingency was selected to provide an additional measure of the degree or strength of the relationship.

CHI-SQUARE-GOODNESS-OF-FIT-TEST

One of the primary objectives of these studies involved the analysis of the distributional form of observed performance in comparison with the distributional form predicted by biorhythms. If an event was equally likely to occur anywhere within a given time-span, the distribution of numerous events during that time-span should be uniform. It follows that if the equally likely hypothesis does not hold, the distribution of events would be something other than uniform. The Chi-Square-Goodness-of-Fit-Test was selected to test the uniformity hypothesis.

In addition, if the uniformity hypothesis can be rejected a technique should be included which will provide a means of identifying the distributional form of observed data. The method of moments and the method of maximum likelihood were used to estimate the parameters of the hypothesized distribution (Kendall and Stewart[14]). Once the parameters were estimated, the hypothesized distribution was again tested using the Chi-Square-Goodness-of-Fit-Test.

The statistical methods discussed above were considered the primary statistical procedures. However, in situations in which the sample size may be small, the Chi-Square-Contingency-Test's expected value criterion may not be valid. If this situation is suspected, the Fischer Test for Exact Probability (without Tocher's modification) should be used. Further, in cases in which the Chi-Square-Goodness-of-Fit-Test's expected criterion is not supported, and the category combination reduces the degrees of freedom to an undesirable level, the Kolmogorov-Smirnov test for distributional form provides an alternative procedure.

Detailed discussion of all of the statistical procedures mentioned above can be found in Siegel.[13]

CRITICALITY

As the concept of criticality occupies a central position in biorhythm theory, the development of a statistical procedure which will allow for the determination of the influence of criticality on performance was considered essential. The approach taken here involved the formulation of probabilistic arguments and the incorporation of the probabilities into a suitable statistical model.

The procedure required the determination of the frequencies of critical days based on the periodicity of the oscillations. A summary of critical day frequencies is presented in Table 6.1. If the events are equally likely to occur at any point in time, the probabilities indicate the likelihood that an observed event will occur on a biorhythmically predictable day. (See Table 6.2.)

Biorhythm data can usually be dichotomized into those days on which critical events occurred and those days on which

Table 6.1 Frequency of Occurrence of Critical Days

CRITICALITY	*BIORHYTHM*	*FREQUENCY (DAYS)*
Single	Physical (P)	11.50
Single	Emotional (E)	14.00
Single	Intellectual (I)	16.50
Double	P X E	161.00
Double	P X I	189.75
Double	E X I	231.00
Triple	P X E X I	2656.50

Table 6.2 Chance Probability of Occurrence of Criticalities

CRITICAL DAY	*0 HOUR VARIANCE*	*±6 HOUR VARIANCE*	*±12 HOUR VARIANCE*
Physical Critical Day (P)	.0869	.1304	.1739
Emotional Critical Day (E)	.0714	.1071	.1428
Intellectual Critical Day (I)	.0606	.0909	.1212
P X E	.0062	.0093	.0124
P X I	.0053	.0078	.0105
E X I	.0043	.0065	.0086

critical events occurred and those days on which critical events did not occur. Therefore, one approach to the question of criticality involves the analysis of the statistical significance of the difference between the number of observed critical event days, and the number that would be expected if critical events occurred by pure chance. In the studies of this chapter, the Binomial model was selected to investigate the statistical significance of this difference in cases involving single criticality. If, however, the interest is in multiple criticality, the Multinomial model should be used.

DATA COLLECTION

Application of the model discussed above included academic performance data collected from four graduate students in the Operations Research Curriculum at the Naval Postgraduate School. Data collected covered 15 courses over a 14 month period. In an attempt to minimize subjectivity, only courses with a highly quantitative content were selected for analysis (e.g., multivariable calculus, operations analysis, linear algebra, linear programming, stochastic models, etc.). The procedure required that a performance scale be developed which would indicate an individual's level of performance on a given examination in a given course. The procedure used involved categorizing performance as (a) "well above average," (b) "above average," (c) "average," (d) "below average," (e) "well below average," and (f) "critical" performance. For example, a student with an average of 75 who, on a given examination scored 90 out of a possible 100 points while the class average at the time of the examination was 80, was considered "well above" average. Subjects were required to rate the level of difficulty or challenge a particular examination represented for that individual. The rating scale for challenge was similar to that used in classifying performance levels.

Admittedly, there is considerable subjectivity involved in the academic data. However, the procedure did provide sufficient information to enable the investigators to categorize the performance data, and provided information on the relative degree of challenge the task presented to an individual student.

In addition to the academic performance data, industrial accident report forms were obtained from a forest products company. The accident report forms received were carefully screened in an attempt to eliminate accidents in which second parties were involved, and/or accidents in which equipment malfunctions or system induced errors contributed to the accident. Therefore, accidents selected for examination included only those accidents involving a single operator and were the result of errors directly attributable to the operator. This procedure yielded a total of 66 accidents for analysis.

RESULTS

ACADEMIC PERFORMANCE

The analysis of the academic performance data involved an examination of observed performance in relation to the intellectual cycle of the theory of biorhythms. The first step in the analysis was concerned with an examination of the possible influence of the intellectual cycle on "good" and "poor" performance. The technique involved the use of the Chi-Square-Contingency-Test.

CHI-SQUARE-CONTINGENCY-TEST OF "GOOD" AND "POOR" ACADEMIC PERFORMANCE

The hypothesis to be tested was that observed "good" and "poor" performance days were related to the intellectual biorhythm cycle. Table 6.3 presents a breakdown of the categorization procedure and the results of the Chi-Square-Contingency-Test. The results suggest that H_0 should be rejected in favor of H_1 ($p < .001$). This indicates that a statistically significant relationship existed between the intellectual biorhythm and academic performance as measured here.

Table 6.3 Chi-Square-Contingency-Test of the Relationship of Academic Performance to the Intellectual Cycle

CYCLE	*OBSERVED "GOOD" ACADEMIC PERFORMANCE*	*OBSERVED "POOR" ACADEMIC PERFORMANCE*
First Quarter of Positive Cycle	21	4
Second Quarter of Positive Cycle	15	9
First Quarter of Negative Cycle	8	14
Second Quarter of Negative Cycle	2	16
	λ^2_3 (.999) = 16.27	V = 25.7*

*p < .001

CHI-SQUARE-COEFFICIENT-OF-CONTINGENCY

The next aspect of the analysis involved an examination of the strength or degree of relationship as measured by the Chi-Square-Coefficient-of-Contingency. The analysis produced a C of .48. When the sample size involved in the analysis is taken into consideration, a C of .48 provides additional evidence of a statistically significant relationship between academic performance and the intellectual biorhythm.

ANALYSIS OF "AVERAGE" PERFORMANCE

As a result of the importance of the concept of challenge in the present interpretation of the theory, and the necessity of identifying "average" performance as a basis for categorizing "above" and "below average" performance, an analysis of the distribution of "average" academic performance days was included.

CHI-SQUARE-GOODNESS-OF-FIT-TEST OF "AVERAGE" PERFORMANCE

The hypothesis to be tested was that the distribution of "average" performance days was not uniformly distributed over the intellectual cycle. Table 6.4 presents the results of the Chi-Square-Goodness-of-Fit-Test for uniformity of "average" academic performance. With three degrees of freedom it was not possible to reject the null hypothesis ($p < .30$). Therefore, based on the above, the uniformity hypothesis of "average" performance data must be accepted.

A Kolmogorov-Smirnov Test for distributional form was applied to the same data in an attempt to provide convergent evidence. A calculated D of .1248 was not statistically significant ($p < .20$). Therefore, as was true for the Chi-Square-Goodness-of-Fit-Test, the null hypothesis of uniformity could not be rejected using the Kolmogorov-Smirnov Test. Thus, it can be assumed that "average" academic performance was equally likely to occur anywhere within the intellectual cycle.

Table 6.4 Chi-Square-Goodness-of-Fit-Test for Uniformity of "Average" Academic Performance

INTERVAL (DAYS)	*OBSERVED FREQUENCY*	$\dot{P}_i$	$n\dot{P}_i$
$1 \leq t \leq 8$	5	.242	6.06
$8 < t \leq 16$	4	.242	6.06
$16 < t \leq 26$	8	.303	7.57
$26 < t \leq 33$	8	.212	5.30

λ_3^2 (.70) = 1.42 $\qquad$ V = 2.28*

$\dot{P}_i$ = probability that an observed event will occur under the null hypothesis of uniformity, during the associated interval

$n\dot{P}_i$ = expected number of observations under the null hypothesis of uniformity

*$p < .30$

ANALYSIS OF "ABOVE AVERAGE" ACADEMIC PERFORMANCE

In the analysis of "above average" academic performance, the categories "well above average" and "above average" were combined.

CHI-SQUARE GOODNESS-OF-FIT-TEST FOR UNIFORMITY OF "ABOVE AVERAGE" ACADEMIC PERFORMANCE

The first analysis involved the uniformity question and tested the hypothesis that the distribution of "above average" academic performance was not uniformly distributed over the intellectual cycle. The results are presented in Table 6.5. The findings indicate that the null hypothesis should be rejected ($p < .001$). Therefore, "above average" academic performance was not equally likely to occur anywhere over the intellectual biorhythm.

A Kolmogorov-Smirnov test for distributional form was performed on the same data with the same hypothesis. A calculated D of .3504 supported the rejection of the null hypothesis of uniformity at the .01 level. Therefore, it would appear safe to conclude that "above average" academic performance was not equally likely to occur anywhere within the intellectual biorhythm.

Table 6.5 Chi-Square-Goodness-of-Fit-Test for Uniformity of "Above Average" Academic Performance

INTERVAL (DAYS)	*OBSERVED FREQUENCY*	$\dot{P}_i$	$n\dot{P}_i$
$1 \leq t \leq 5$	13	.151	6.96
$5 < t \leq 10$	10	.151	6.96
$10 < t \leq 15$	13	.151	6.96
$15 < t \leq 20$	4	.151	6.96
$20 < t \leq 25$	4	.151	6.96
$25 < t \leq 33$	2	.242	11.20
	λ_5^2 (.999) = 20.52		V = 21.88*

$\dot{P}_i$ = probability that an observed event will occur under the null hypothesis of uniformity during the associated interval

$n\dot{P}_i$ = expected number of observations under the null hypothesis of uniformity

*$p < .001$

In view of the findings discussed above, it was considered essential that an attempt be made to define the distributional form of observed "above average" academic performance. This involved plotting the observations and estimating the parameters of the hypothesized distribution. The techniques used for parameter estimation were the method of moments and the method of maximum likelihood (Kendall and Stewart[14]).

CHI-SQUARE-GOODNESS-OF-FIT-TEST OF THE ESTIMATED DISTRIBUTION OF "ABOVE AVERAGE" ACADEMIC PERFORMANCE

Using the procedures mentioned above, the hypothesized distributional form was estimated to be normal with a mean of 8.5 and a standard deviation of 5. The hypothesis to be tested was that the distributional form of "above average" academic performance was other than N(8.5, 5). The results of the Chi Square-Goodness-of-Fit-Test are presented in Table 6.6. The findings indicate that the null hypothesis could not be rejected ($p < .50$). Therefore, the distributional form of "above average" academic performance can be considered as normal with a mean of 8.5 and a standard deviation of 5.

Table 6.6 Chi-Square-Goodness-of-Fit-Test for the Estimated Distributional Form N(8.5, 5) of "Above Average" Academic Performance

INTERVAL (DAYS)	*OBSERVED FREQUENCY*	$\dot{P}_i$	$n\dot{P}_i$
$-\infty \leq t \leq 3$	7	.136	6.24
$3 < t \leq 12$	24	.623	28.64
$12 < t \leq -\infty$	15	.242	11.12
	λ_2^2 (.50) = 1.386		V = 2.19*

$\dot{P}_i$ = probability that an observed event will occur under the null hypothesis of N(8.5, 5) during the associated interval

$n\dot{P}_i$ = expected number of observations under the null hypothesis of N(8.5, 5)

*$p < .50$

ANALYSIS OF "BELOW AVERAGE" ACADEMIC PERFORMANCE

To strengthen the analysis the categories of "well below average" and "below average" performance were combined.

CHI-SQUARE-GOODNESS-OF-FIT-TEST FOR UNIFORMITY OF "BELOW AVERAGE" PERFORMANCE

The first analysis involved the question of uniformity in the distributional form of "below average" performance. The hypothesis to be tested was that the distribution of "below average" academic performance was not uniformly distributed over the intellectual cycle.

The results of the Chi-Square-Goodness-of-Fit-Test are presented in Table 6.7. The findings suggest that the null hypothesis of uniformity could not be rejected at a level of significance greater than $p < .10$.

Table 6.7 Chi-Square-Goodness-of-Fit-Test for Uniformity of "Below Average" Academic Performance

INTERVAL (DAYS)	OBSERVED FREQUENCY	$\dot{P}_i$	$n\dot{P}_i$
$1 \leq t \leq 8$	4	.242	10.42
$8 < t \leq 16$	9	.242	10.42
$16 < t \leq 27$	21	.333	14.34
$27 < t \leq 33$	9	.182	7.80
	λ_3^2 (.90) = 6.25		V = 7.42*

$\dot{P}_i$ = probability that an observed event will occur under the null hypothesis of uniformity during the associated interval

$n\dot{P}_i$ = expected number of observations under the null hypothesis of uniformity

*$p < .10$

A Kolmogorov-Smirnov Test on the same data testing the same hypothesis produced similar results. A calculated D of .1825 indicated that the null hypothesis could not be rejected ($p < .15$).

While the hypothesis of nonuniformity could not be accepted at the desired level of confidence, there were indications that the distributional form of "below average" academic performance data was more complex than uniform. Therefore, an attempt was made to determine the distributional form of the data. The procedure used was similar to that described for "above average" academic performance data. The results of the estimating techniques suggested a Beta distribution with parameters of 4 and 2.

CHI-SQUARE-GOODNESS-OF-FIT-TEST OF THE ESTIMATED DISTRIBUTION OF "BELOW AVERAGE" ACADEMIC PERFORMANCE

The hypothesis to be tested by the Chi-Square-Goodness-of-Fit-Test was that the distributional form of "below average" academic performance was other than $\beta(4.2)$. The results indicated that the null hypothesis could not be rejected ($p < .30$). (See Table 6.8.)

Table 6.8 Chi-Square-Goodness-of-Fit-Test for the Estimated Distributional Form of "Below Average" Academic Performance

INTERVAL (DAYS)	*OBSERVED FREQUENCY*	$\dot{P}_i$	$n\dot{P}_i$
$1 \leq t \leq 20$	13	.350	15.05
$20 < t \leq 30$	24	.575	24.73
$30 < t \leq 33$	6	.075	3.23
	λ_2^2 (.70) = 2.41		V = 2.69*

$\dot{P}_i$ = probability that an observed event will occur under the null hypothesis $\beta(4, 2)$ during the associated interval.

$n\dot{P}_i$ = expected number of observations under the null hypothesis of $\beta(4, 2)$.

*$p < .30$

The findings presented above suggest that the distributional form of "below average" academic performance was more complex than observed for "average" or "above average" academic performance.

INDUSTRIAL ACCIDENTS

Industrial accident data was analyzed in terms of a possible relationship between the events (accidents) and the physical biorhythm. The data to be analyzed was arranged in tabular form depicting accident frequency in terms of the day of the physical cycle. Since the observed accidents must be considered as falling in a "below average" performance category, testing for the presence of a physical biorhythm influence must begin with a test for uniformity.

CHI-SQUARE-GOODNESS-OF-FIT-TEST FOR UNIFORMITY OF INDUSTRIAL ACCIDENTS

The hypothesis tested was that the occurrence of industrial accidents was not uniformly distributed over the physical cycle. The results of the analysis are presented in Table 6.9. The null hypothesis could not be rejected at the .05 level, but could be rejected at the .10 level. The results of a Kolmogorov-Smirnov Test produced a calculated D of .1620 which was significant at the .10 level. (See Table 6.9.)

Table 6.9 Chi-Square-Goodness-of-Fit-Test for Uniformity of Industrial Accidents

INTERVAL (DAYS)	*OBSERVED FREQUENCY*	$\dot{P}_i$	$n\dot{P}_i$
$1 \leq t \leq 5$	10	.217	14.35
$5 < t \leq 10$	8	.217	14.35
$10 < t \leq 15$	20	.217	14.35
$15 < t \leq 20$	15	.217	14.35
$20 < t \leq 23$	13	.130	8.60
	λ_4^2 (.90) = 7.77		V = 8.63*

$\dot{P}_i$ = probability that an observed event will occur under the null hypothesis of uniformity during the associated interval.

$n\dot{P}_i$ = expected number of observations under the null hypothesis of uniformity

*$p < .10$

A note from

DR. MAURICE A.E. ROTHMAN

However, as a result of the theory having a built-in variance of ±12 hours due to the uncertainty of the hour of birth, it was decided that the data be shifted to account for this variance, and the data re-analyzed for distributional form. By shifting the data by ±12 hours it was possible to reject the null hypothesis at the .001 level. (See Table 6.10.) Results of a Kolmogorov-Smirnov test on the same data yielded a calculated D of .2378 which was significant at the .01 level.

Table 6.10 Chi-Square-Goodness-of-Fit-Test for Uniformity of Industrial Accidents After Shifting Data ±12 Hours

INTERVAL (DAYS)	*OBSERVED FREQUENCY*	$\dot{P}_i$	$n\dot{P}_i$
$1 \leq t \leq 5$	6	.217	14.35
$5 < t \leq 10$	7	.217	14.35
$10 < t \leq 15$	22	.217	14.35
$15 < t \leq 20$	15	.217	14.35
$20 < t \leq 23$	16	.130	8.60
	λ_4^2 (.999) = 18.46		V = 19.71*

$\dot{P}_i$ = probability that an observed event will occur under the null hypothesis of uniformity during the associated interval

$n\dot{P}_i$ = expected number of observations under the null hypothesis of uniformity

*$p < .001$

Therefore, shifting the data to take into consideration the uncertainty of the hour of birth enabled the rejection of the uniformity hypothesis with a much higher degree of confidence. However, it should be emphasized that without shifting the data, the results were suggestive of the possibility that accidents were not equally likely to occur anywhere within the physical cycle.

ANALYSIS OF LOST-TIME VS NON-LOST-TIME ACCIDENTS

The data was dichotomized into lost-time accidents and non-lost-time accidents. A lost-time accident was defined as an accident in which an employee was forced to remain off the job for 24 hours. A non-lost-time accident involved an incident which was reported but did not result in time away from the task.

CHI-SQUARE-CONTINGENCY-TEST FOR UNIFORMITY OF LOST-TIME AND NON-LOST-TIME INDUSTRIAL ACCIDENTS

A Chi-Square-Contingency-Test was performed on the data in an attempt to analyze the relationship between the accident categories and the physical biorhythm cycle. The hypothesis to be tested was that lost-time and non-lost-time industrial accidents were not uniformly distributed across the physical biorhythm. The results of the categorization procedure and the Contingency Test analysis are presented in Table 6.11. Based on the results it was possible to reject the hypothesis that the accidents were equally likely to occur at any point within the physical cycle ($p < .05$).

Table 6.11 Chi-Square-Contingency-Test of the Relationship of Lost-Time and Non-Lost-Time Accidents to the Physical Biorhythm

CYCLE	LOST-TIME ACCIDENTS	NON-LOST-TIME ACCIDENTS
Positive Portion of the Cycle	1	21
Negative Portion of the Cycle	11	33
	λ_1^2 (.95) = 3.84	V = 4.18*

*p < .05

CHI-SQUARE-COEFFICIENT-OF-CONTINGENCY

Application of Coefficient-of-Contingency Test to the data provided an indication of the degree of the relationship between accidents and the physical biorhythm. The results of the test yielded a C of .244. It should be emphasized that a C of .244 provides additional evidence of a statistically significant relationship between industrial accidents and the physical biorhythm.

CRITICALITY

Because only one critical event was identified in the academic performance data, it was not possible to include a criticality analysis. In the industrial accident data, however, it was possible to consider lost-time accidents as "critical" events and examine the relationship of such events to "physical critical days."

The first step in the analysis required the determination of the probabilities associated with the occurrence of a physical critical day. The calculated probability of a physical critical day was .1739.

By analyzing the 11 lost-time accidents which occurred during the negative portion of the physical cycle and by taking into consideration the ±12 hour variance it was determined that the expected (mean) number of criticalities was 1.91 with a variance of 1.58 and a standard deviation of 1.25. Four lost-time accidents occurred on critical days. Based on the calculated probabilities, it can be assumed that the occurrence of the observed four criticalities was a rare event. Figure 6.2 illustrates that 90% of the probability of occurrence of critical observations is contained within 1.5 standard deviations of the mean.

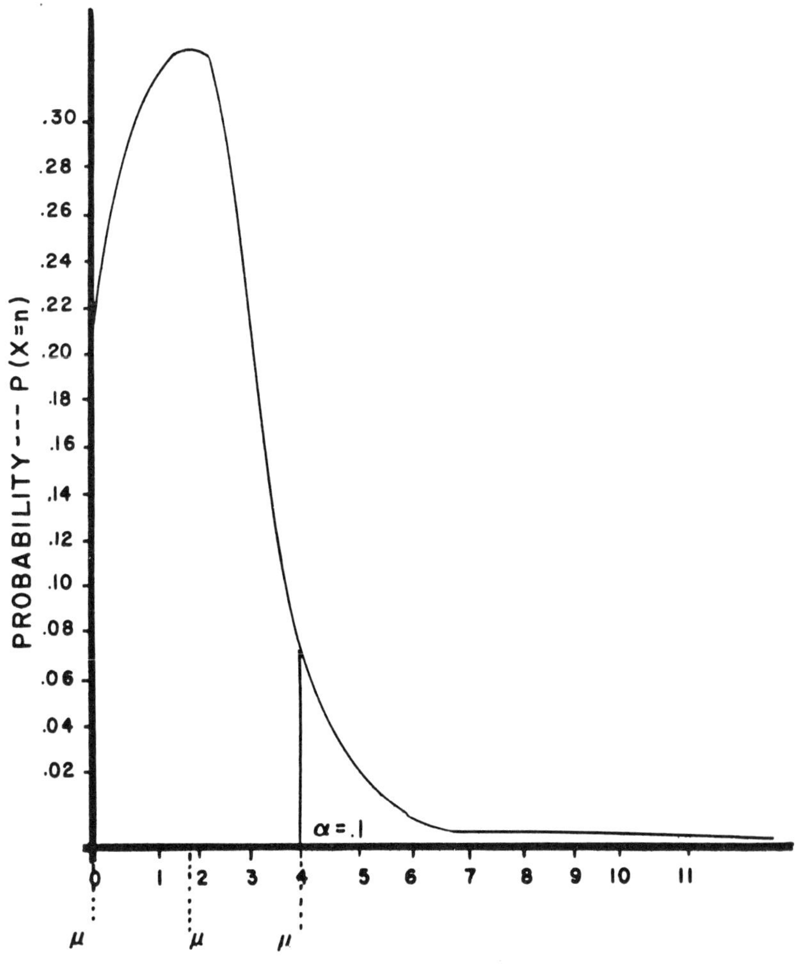

Figure 6.2 Probability of Criticality

DISCUSSION

The development of a statistical model for the study of biorhythms is dependent upon the researcher's interpretation of the theory. In the past, the majority of the research effort has been devoted to the investigation of the concept of criticality, and has neglected the analysis of the total theoretical rhythmicity associated with the theory of biorhythms. Therefore, the attempt here has been to develop a combined statistical model which would provide a method for analyzing the total theory in terms of biorhythmically predictable oscillations in human performance.

The analyses of academic performance and industrial accident data cannot be considered as providing a validation of the theory of biorhythms. However, the analyses do represent illustrations of the application of the statistical model and are suggestive of a possible biorhythmic influence on human performance.

Analyses of academic performance provided an example of the application of the combined statistical model for the investigation of biorhythms. The findings suggested a relationship between performance levels and the positive and negative phases of the intellectual cycle. That is, "good" academic performance occurred with greater frequency during the positive phase of the curve and "poor" academic performance occurred with greater frequency during the negative phase of the cycle.

Based on the present interpretation of the theory, an examination of "average" academic performance revealed a uniform distribution and indicated that "average" academic performance was equally likely to occur at any point in the cycle. These findings were interpreted as lending support to the hypothesis that biorhythms represent a variable in performance potential which requires a challenging situation in order to be demonstrated. Therefore the observed uniformity of "average" performance may be interpreted as the result of a task which was insufficiently demanding to prevent masking of the biorhythmic influence by participants.

In terms of "above average" academic performance, the hypothesis of uniform distribution was rejected at the .001 level. Thus, in contrast to "average" academic performance, "above average" academic performance was not equally likely to occur throughout the intellectual cycle. Therefore, in line with the present interpretation of the theory, it can be postulated that "above average" academic performance represented a greater degree of challenge. As such, biorhythmic influence could not be easily masked by the participants. The results were interpreted as support for the notion that there was a significant impact of biorhythms upon "above average" academic performance.

Analysis of "below average" academic performance did not reveal the same characteristics of either "average" or "above average" performance. Considering "below average" academic performance, the uniformity hypothesis could not be rejected with the same confidence as was true of "above average" performance. That is, the hypothesis of uniformity could only be rejected at the .10 level. The procedures used in estimating the distributional form of "below average" academic performance revealed characteristics which differed significantly from those with "above average" performance in that a complex Beta distribution was identified [$\beta(4,2)$]. These findings were interpreted as suggesting that "poor" performance may be the result of the interaction of numerous variables, including oscillating physical, emotional and intellectual subsystems. Therefore, "poor" performance may represent a more "complex" form of behavior than either "average" or "above average" performance. Indeed, "poor" performance may be the outcome of interference of a "feedback subsystem." Perhaps at the time "below average" academic performance occurred, there was a biological condition of inadequate intellectual capacity. The intellectual subsystem might have had a poor predisposition toward confronting intellectual problems. The accident data provided additional support for the notion of a biological predisposition in that the distributional form and time of occurrence of "below average" performance was related to the negative phase of the physical cycle. Lost-time accidents and non-lost-time accidents were not found to be equally likely ($p < .05$). The results presented striking confirmation of the uneven distribution of lost-time accidents. Only one lost-time accident occurred in the positive phase of the physical cycle while 11 occurred in the negative phase. If the negative phase of the physical biorhythm is considered to be a period during which there is an alteration of either sensory thresholds and/or physical adaptability, then it may follow that more severe accidents (lost-time) will occur during this phase. The results indicated that one-third more non-lost-time accidents occurred in the negative phase (33) of the cycle than in the positive phase (21). This seems to add further support for an accident period proneness or biological predisposition. The results also showed that far more non-lost-time accidents (21) than lost-time accidents (1) occurred in the positive phase of the physical cycle. Other factors than biological ones are obviously operating during any performance or activity. It can be postulated that at times less serious accidents may occur even during the positive physical phase. While in the negative phase the balance is tipped so that the body finds itself less able to cope effectively with its environment. Hence, we find more accidents and more serious accidents occurring during the negative physical phase in which it is postulated there is less than optimal physical functioning.

As a result of considering accidents as examples of "poor" performance, the combined statistical model could not be used in the analysis of industrial accidents. However the results did suggest that accidents were not uniformly distributed over the physical cycle ($p < .10$). In addition, by shifting the data to take into consideration the uncertainty of the hour of birth, the uniformity (equally likely) hypothesis could be rejected at a much higher level of confidence ($p < .001$). These results were interpreted as suggesting a possible biorhythmic influence in accident occurrence. Further, when the data was dichotomized into lost-time and non-lost-time accidents, a strong relationship between accidents and the physical biorhythm was indicated ($p < .05$).

As the stated objective of this work was to analyze the complete cycle as opposed to concentrating on criticality, an examination of critical days was not emphasized. Further, the academic performance did not provide sufficient data for a criticality analysis. The accident data, however, was analyzed for the presence of a criticality phenomena. Under the null hypothesis of pure chance, an average of 1.91 accidents with a variance of 1.58 and a standard deviation of 1.25 would be expected. The fact that four accidents were observed to occur on physical critical days must be considered as a rare event. The data indicated that there was not only broad negative periodicity but also specific critical days on which a statistically significant number of lost-time accidents occurred. The suggestion here is that a definite accident-critical day relationship did exist in the sample.

In general, the theory of biorhythms has been considered an esoteric subject with its origins in the occult sciences. The emergence and seeming victory of Behaviorism in America destroyed, or at least minimized, any emphasis upon the inherent patterns of the organism. This study represents an attempt to apply accepted statistical techniques to the study of the theory of biorhythms. The results obtained were suggestive and indicate the need to pursue the scientific investigation of the concept. At the present time, numerous laboratory and field studies are being conducted at the Man-Machine Systems Design Laboratory at the Naval Postgraduate School in an attempt to verify the presence or absence of isochronic phenomena associated with the theory. Research is also pursuing the questions of the origins and nature of the oscillations in terms of cycle interactions, cycle modifications and physiological/psychological correlates. The results of such experimentation would have significant ramifications for many fields including safety and basic human functioning.

REFERENCES

1. Colquhoun, W. P., Preface. In W. P. Colquhoun (Ed.), *Biological Rhythms and Human Performance.* New York: Academic Press, 1971, vi-vii.
2. Alluisi, E. A. and Chiles, W. D., "Sustained Performance, Work-Rest Scheduling, and Diurnal Rhythms in Man," *Acta Psychologica,* 1967, 27, 436-442.
3. Wèver, R., "Circadian Rhythms of Some Psychological Functions Under Different Conditions." NATO: AGARD Report No. CP-74-70, Conference Proceedings on Rest and Activity Cycles for the Maintenance of Efficiency of Personnel Concerned with Military Flight Operations, 1970, (AD 717 265).
4. Hauty, G. T., "Relationships Between Operator Proficiency and Effected Changes in Biological Periodicity." *Aerospace Medicine,* 1963, 34(2), 100-105.
5. Hauty, G. T. and Adams, T., "Phase Shifts of the Human Circadian System and Performance Deficit During the Periods of Transition: II. West-East Flight." *Aerospace Medicine,* 1966, 37(10), 1027-1033. (a).
6. Hauty, G. T. and Adams, T., "Phase Shifts of Human Circadian System and Performance Deficit During the Periods of Transition: III. North-South Flight." *Aerospace Medicine,* 1966, 37(12), 1257-1262. (b).
7. Lobban, M. C., "Time, Light and Diurnal Rhythms." In O. G. Edholm and A. L. Bacharach (Eds.), *The Physiology of Human Survival.* New York: Academic Press, 1965, 351-386.

8. Lobban, M. C. and Tredre, B. E., "Renal Diurnal Rhythms in Human Subjects During Bed Rest and Limited Activity," *Journal of Physiology,* 1964, 171 (2), 26-27.
9. Oatley, K. and Goodwin, B. C., "The Explanation and Investigation of Biological Rhythms." In W. P. Colquhoun (Ed.), *Biological Rhythms and Human Performance.* New York: Academic Press, 1971, 1-38.
10. Thommen, G. S., *Is This Your Day?* (2nd Edition) New York: Crown Publishers, 1973.
11. Wallenstein, M. R. and Roberts, N. L., "All Together on Bio-Curve," *Human Behavior,* April 1973, 8-15.
12. Willis, H. R., "Biorhythms and Its Relationship to Human Error," Proceedings of the Sixteenth Annual Meeting of the Human Factors Society, Los Angeles, October 1972, pp. 274-282.
13. Siegel, S., *Nonparametric Statistics.* New York: McGraw-Hill, 1956.
14. Kendall, M. G. and Stewart, A., *The Advanced Theory of Statistics,* Vol. 2 (2nd ed.) New York: Hafner, 1947.

7

Laboratory Investigations for the Analysis of Biorhythms

In Chapter 1, Dr. Douglas E. Neil's laboratory experiments to determine whether a relationship between human performance and biorhythms may be proven were discussed briefly. In this chapter the details of one such laboratory investigation to examine this relationship are presented. The material which follows is taken from the paper, "A Laboratory Investigation of Biorhythms," by Douglas E. Neil and Francis L. Sink of the Man-Machine Systems Design Laboratory, Monterey, California. It is printed with the permission of Aviation, Space and Environmental Medicine *as it appeared in Volume 47, Number 4, page 425. Once again, figure and table numbers have been changed for consistency with those used in this book.*

A LABORATORY INVESTIGATION OF BIORHYTHMS

By Douglas E. Neil and Francis L. Sink

SUMMARY

In this study three subjects were tested on an information processing task on a daily basis for a period of 70 days. Performance measurement included reaction time, movement time and information processing rate. The data set obtained was subjected to Fast Fourier Transforms (FFT) in an attempt to identify periodicities in performance. Twelve significant harmonics were identified and nine were found to be within one day of at least one of the cycles hypothesized by the theory of biorhythms. The results were interpreted as suggesting the possibility of a biorhythmic influence in the performance of the task.

METHODS AND MATERIALS

Three male Naval Officer students at the Naval Postgraduate School served as subjects. Ages ranged from 30 to 36 years. All were volunteers and received no compensation for their participation. Participants had no knowledge of their biorhythms during the course of the examination.

The apparatus consisted of a panel of two rows of displays with four displays in each row, and a corresponding response panel consisting of eight response buttons arranged in two rows of four. Stimuli consisted of the digits one through eight which were displayed on the stimulus panel. Each response button was related to only one display in the stimulus panel. The ordering of the stimulus and response panels was such that response buttons one through four corresponded to the stimulus panels one through four, and panels five through eight with buttons five through eight.

The task required subjects to depress the appropriate response button upon the presentation of a stimulus (digit) in the display panel. Subjects were instructed to respond as quickly as possible upon the presentation of a stimulus. Stimulus rate was one per second with the actual order of presentation randomized, and changed daily. A 10 second warm-up period was allowed each day prior to the start of testing.

Performance measures included reaction time, movement time and information processing rate. Reaction time was directly observable from the apparatus in that reaction time was recorded directly on punched tape and subsequently submitted to a PDP/8E laboratory computer for analysis. Movement time and information processing rate were determined on the basis of "Hick's Law" which suggests that a choice reaction time is a function of the probabilities of stimuli, and as such is an increasing linear function of the amount of information or uncertainty associated with a stimulus.

The algebraic or functional form of "Hick's Law" can be stated as:

$$RT = a + bH \tag{7-1}$$

where:

a = movement time

b = seconds/bit of information (1/b information processing rate)

H = uncertainty associated with the stimuli (for an equally likely stimuli $H = \log_2 n$)

In the present study, three difficulty levels were administered sequentially. The first task involved making a known response to two equally likely stimuli (a one bit task), the second task involved making a known response to four equally likely stimuli (a two bit task), and the third task involved eight equally likely stimuli (a three bit task). It was felt such a procedure would aid in attempting to rule out the possibility of masking out any biorhythm influence by ensuring that the task would be challenging. Each subject continued to perform at each difficulty level until five response times were recorded.

The fifteen daily reaction time scores were used in a linear regression model to produce a least squares error estimate of a and b discussed in equation 7-1. The equations used were as follows:

$$a = \frac{\sum_{i=1}^{15} Y_i}{15} - b\frac{\sum_{i=1}^{15} x_i}{15} \qquad (7\text{-}2)$$

and

$$b = \frac{N\sum_{i=1}^{15} X_i Y_i - \sum_{i=1}^{15} X_i \sum_{i=1}^{15} Y_i}{N\sum_{i=1}^{15} (X_i)^2 - \left(\sum_{i=1}^{15} X_i\right)^2} \qquad (7\text{-}3)$$

where: X_i = bits of information
Y_i = reaction time
N = total number of data points

The first step in the analysis involved taking the average of the 15 reaction times recorded during a daily session and examining the distributional form of performance. The distributional form of the 2 bit information processing task was observed to be approximately normal (μ, α^2). The assumption was made that on any day the distribution of the one bit task was normal (μ – c, σ^2) and the distribution of the three bit task was normal (μ + c, σ^2), where c represents a constant which consists of the inverse of information processing rate. The constant c was added to the one bit task and subtracted from the three bit task data to yield 15 data points per day which were normally distributed (μ, α^2). The mean of the distribution was estimated by the following:

$$\overline{X} = 1/n \sum_{n=1}^{15} X_n \qquad (7\text{-}4)$$

The variance was estimated using the unbiased estimator

$$\sigma^2 = 1/n\text{-}1 \sum_{n=1}^{15} (\overline{X} - X_n)^2 \qquad (7\text{-}5)$$

where, in both equations

n = number of data points
X_n = value of the nth data point

With the estimated value of σ^2, the confidence interval for α–.05 was calculated using the Student distribution. The relation used was

$$L = 2t_{n\text{-}1}\,(1 - \alpha/2)\sqrt{(S^2/n)} \qquad (7\text{-}6)$$

where $n-1$ = degree of freedom
n = number of data points
$(1-\alpha/2)$ = argument for t distribution

This procedure yielded the confidence intervals centered around X. (Table 7.1.)

Table 7.1 Confidence Intervals Centered Around the Mean

SUBJECT	*CONFIDENCE (L)*
One	.036
Two	.052
Three	.044

The grand mean (X_G) was computed for each subject and the data analyzed to determine if any data points were outside the interval $X_G = \pm L/2$. In all cases the data was observed to lie outside the interval, suggesting a significant variation in daily reaction time performance. This observation lead to the application of Fourier Analysis of the data. In addition to reaction time, movement time and information processing rate data were subjected to Fourier Analysis.

The FFT Analysis indicated periodicities, and the function described by the data set was of the following form:

$$Y(NN) = \frac{a_o}{2} + \sum_{n=1}^{35} \left(a_n \operatorname{Cos} \frac{\Pi(NN)^P}{N} + b_n \operatorname{Sin} \frac{\Pi(NN)^P}{N} \right) \quad (7\text{-}7)$$

where $Y(NN)$ = value of function on day NN
N = ½ (number of data points)
NN = 0, 1, 2 . . . 2N-1 (days)
P = harmonic number
a_n = Fourier coefficient
b_n = Fourier coefficient

An existing subroutine at the Naval Postgraduate School Computer Library was used to compute the coefficients a_n and b_n. The root mean square value of a_n and b_n was also computed to give the amplitude of the nth harmonic. In order to get a better approximation of the frequency envelope at the lower frequencies, the data set was reduced in increments of one data point until the final data set consisted of 46 points. This operation is valid under the assumption that the amplitude of the various frequency components do not change over time.

The last step in the frequency analysis of the data involved the levels of significance of the harmonics. Under the null hypothesis that f(t) was a Gaussian Process, the coefficients of the sine and cosine functions in the Fourier series can be assumed to be independent, identically distributed Normal with a mean of zero and a variance of α^2. Since

$$A_n^2 = a_n^2 + b_n^2 \quad (7\text{-}8)$$

where A_n = amplitude of the nth harmonic

and

$$A_n^2 \sim \exp\left(\frac{1}{2\sigma^2}\right) \quad (7\text{-}9)$$

therefore, σ^2 can be estimated by the unbiased estimator

$$S^2 = \frac{1}{N-1} \sum_{n=1}^{N} (\mu - a_n)^2 \qquad (7\text{-}10)$$

where N = number of observations of a_n.

Since μ was assumed to be zero, this reduces to

$$S^2 = \frac{1}{N-1} \sum_{n=1}^{N} (a_n)^2 \qquad (7\text{-}11)$$

The probability (α) that an amplitude greater than A_n would arise from a Gaussian Process was found from the relation

$$\alpha = e^{-A_n^2/2S^2} \qquad (7\text{-}12)$$

RESULTS

The results of the Fourier analysis of reaction time data are presented in Table 7.2. A close examination of the information presented in Table 7.2 suggests considerable individual variation in performance cycling. As such, subject number one did not manifest rhythmicities which could be considered statistically significant. Subject number two's reaction time performance was observed to cycle with periods of 11.0 and 15.0 days at levels of statistical significance of $P < .02$ and $P < .001$ respectively. Further, the observed periodicities were within one-half day of the physical biorhythm (physical cycle period = 11.5 days) and one day of the emotional biorhythm (emotional cycle period = 14 days).

Table 7.2 Fourier Analysis of Reaction Time Data

PERIOD (Days)	S^2	*AMPLITUDE*	α	*PHASE (Degrees)*
		(subject one)		
10	13.3×10^{-6}	.0113	.061	177°
14.5	13.3×10^{-6}	.0085	.384	325°
		(subject two)		
11	11.7×10^{-6}	.0115	.020*	91°
15.0	11.7×10^{-6}	.0137	.001*	150°
		(subject three)		
14.8	16.4×10^{-6}	.0149	.007*	16°
35	16.4×10^{-6}	.0158	.003*	328°

*P < .05

Subject number three's reaction time performance was also observed to cycle during the testing period with significant periods of 14.8 and 35 days. The 14.8 day period is a close approximation of the 14 day emotional cycle periodicity. However, the 35 day period does not relate to any of the periods hypothesized by the theory.

Table 7.3 presents the results of a similar analysis of movement time data recorded during the experiment. As was true of the reaction time data, subject number one's performance was not characterized by cycles with periods approximating the theoretical periodicity predicted by the theory of biorhythms. However, subject one's performance did reveal a highly significant rhythmicity (P < .0007) with a period of 30 days. Subject number two's movement time performance, however, did appear to cycle with a period of 11.5 days. This rhythmicity was highly significant (P < .02) and was in precise agreement with the physical biorhythm cycle. The performance of the third subject was also observed to cycle with periods of 11.8

Table 7.3 Fourier Analysis of Movement Time Data

PERIOD (Days)	S^2	*AMPLITUDE*	α	*PHASE (Degrees)*
		(subject one)		
19	33.8×10^{-6}	.0175	.074	131°
30	33.8×10^{-6}	.0265	.0007*	175°
		(subject two)		
16.7	81.6×10^{-6}	.0195	.518	116°
11.5	81.6×10^{-6}	.0307	.021*	3°
		(subject three)		
11.8	55.9×10^{-6}	.0304	.001*	58°
16.25	55.9×10^{-6}	.0312	.001*	145°

*$P < .05$

and 16.25 days, both of which were significant at the .001 level. These periods are in close agreement with the physical and the intellectual (period = 16.5 days) cycles.

The results of the analysis of information processing rate is presented in Table 7.4. As was true in the previous performance measures, subject number one's performance while manifesting a significant rhythmicity ($P < .0007$), with a period of 29 days, differed from the remaining subjects and from the theoretical predictions. Subject two's information processing rate was observed to oscillate rhythmically with a period of 11 days.

Once again, the observed periodicity is in close agreement with that predicted by the physical cycle of biorhythms. Subject number three's information processing performance manifested two cycles with periods of 11.5 and 16 days and levels of statistical significance of $P < .007$ and $P < .003$ respectively. The 11.5 day periodicity duplicates the physical cycle precisely,

Table 7.4 Fourier Analysis of Information Processing Rate

PERIOD (Days)	S^2	AMPLITUDE	α	PHASE (Degrees)
		(subject one)		
29	8.55	12.8	.0007*	357°
19	3.55	6.68	.42	155°
		(subject two)		
11	.1755	1.43	.027*	160°
23	.1755	1.03	.297	144°
14	.1755	.98	.375	270°
16.5	.1755	.98	.375	53°
		(subject three)		
11.5	.634	2.94	.007*	67°
16	.634	3.12	.003*	178°

*$P < .05$

and the 16 day period is within one-half day of the periodicity predicted by the intellectual cycle.

As a result of the frequency with which periods within one day of 11.5, 14 and 16.5 were observed, a decision was made to attempt to determine the possibility of a variable(s) in the experimental setting giving rise to the observed periodicity. The approach taken involved the determination of the phase associated with each cycle. Tables 7.2, 7.3, and 7.4 present the phases associated with each observed periodicity. The findings were interpreted as indicating that variables present in the actual experiment were not responsible for inducing the observed cyclic phenomena. The basis for this conclusion involves the fact all subjects performed the same task during each test session. Therefore, if conditions present in the experimental setting were responsible for the periodicity, such should have been revealed by the phase angles in the data.

DISCUSSION

Numerous researchers using field data, have provided evidence in support of biological cycling approximating the periodicity suggested by the theory of biorhythms. Many difficulties surround the use of such data, including (a) the data generally has concentrated on negative events (e.g. accidents), (b) field data is generally restricted to a single data point for each individual, and (c) field investigations of the type undertaken do not generally provide for sufficient control to allow for the adequate definition of the theory and the concept it purportedly represents.

In this study, an attempt was made to examine the theory under controlled laboratory conditions. The results were interpreted as indicating that the performance measures used cycled with a considerable measure of agreement with the periodicity predicted by biorhythms. However, the obvious exception was the performance of subject number one. This subject's performance was not observed to cycle in agreement with the theory, or in the same pattern as his fellow subjects. As a result of the observed differences, it can be postulated that (a) the theory is invalid and that hypothesized cycles are nonexistent, or (b) that considerable individual variations exist in terms of the influence of the hypothesized rhythmicity on a given individual's performance capacity. Assuming that alternative (b) is correct, it may be that a subject can effectively mask out the influence of the cycles on performance by "trying harder," in much the same way McFarland[2] has suggested frequently occurs in laboratory research involving fatigue. As such, Neil[3] has suggested that the concept of challenge is of prime importance in laboratory studies of biorhythms. In situations in which the task required of subjects is insufficient in terms of the demands placed on participants, the influence of biorhythms can be effectively masked by the subject's trying harder. Therefore, it may be, that in the present study the task which confronted the subjects did not represent a sufficient challenge to subject number one, and as a result, performance was characterized by an absence of periodicity.

CONCLUSIONS

As indicated above, the results of the present pilot laboratory investigation of biorhythms were suggestive of a periodicity in performance which in many respects corresponded with the oscillations predicted by the theory. However, the nagging question regarding the validity of the concept persists, primarily as a result of the lack of firm theoretical foundations, and the rather dubious nature of much of the research used as evidence of its existence. However, when all the evidence is considered, the results are thought provoking and indicate the need to pursue the scientific investigation of the concept. If it should turn out that the theory and the concept it represents are valid, the contribution such knowledge could have in man-machine systems design, safety and management decision-making situations could represent a method whereby human variability could be reduced and man's contribution to system efficiency improved.

REFERENCES

1. Alluisi, E. A., 1970. "Information and Uncertainty: The Metrics of Communication," *Systems Psychology,* K. B. DeGreene (Ed.), New York: McGraw-Hill, pp. 171-206.
2. McFarland, R. A., 1971. "Understanding Fatigue in Modern Life," *Economics,* 14:1-10.
3. Neil, D. E., 1974. "Biorhythms and Industrial Accidents." Paper presented at the 1974 National Safety Congress, Chicago, Ill.
4. Thommen, G. S., 1973. *Is This Your Day?* New York: Crown Publishers.

5. Wallenstein, M. R. and Roberts, Nancy L., 1973. "All Together on the Bio-Curve," *Human Behavior,* April, pp. 8-15.
6. Willis, H. R., 1972. "Biorhythm and Its Relationship to Human Error." Proceedings of the Sixteenth Annual Meeting of the Human Factors Society, Los Angeles, pp. 274-282.

8

Establishing An In-Plant Biorhythm Program

TEST FOR BIORHYTHM CORRELATION

The theory of biorhythms as applied to industrial safety is, to date, an unproven science. The first step in establishing an in-plant biorhythm program involves determining if a correlation exists between employee biorhythms and plant accidents. The conflicting results which have been reported by various investigators indicate that biorhythms may be correlated with certain types of accidents but may not be related in other situations.

Simple analytical tests may be applied to accident data to establish if a correlation with biorhythms potentially exists. If a positive correlation is initially indicated, a detailed analysis would be necessary to establish a definite correlation, and to provide justification for management to undertake a commitment to a biorhythm program.

ACCIDENT CLASSIFICATION

Accidents must be classified by severity and cause in order to perform a meaningful biorhythm analysis. Accidents may be classified into two severity categories:

1. Lost time accidents
2. Non-lost time accidents

The results of Neil's study in Chapter 6 indicate a stronger relationship between biorhythm cycles and lost time accidents. Thus only those lost time accidents of 24 hours or more should be considered in the initial analysis.

Accidents should also be classified by cause:

A. Accidents which result from an individual's actions

B. Accidents which result from another individual's actions

C. Accidents which result from factors beyond the control of any individual.

The simplest test involves analysis of those individuals who fall into category "A." Analysis of accidents involving two individuals, category "B," would involve investigation of both individuals' biorhythms. Category "C" would include accidents such as those related to a poisonous gas leak or explosion. These accidents should not be considered in the initial test.

BIORHYTHM CALCULATIONS

In order to determine the number of "hits" (those accidents which occur on the accident prone days of the biorhythm cycle), it is necessary to analyze the individual's biorhythm. The procedure is to add up the total number of days in the individual's life from the day of birth to the day in question (such as the date of an accident) and to determine the position of the biorhythm cycles on that day. To assist in understanding biorhythm calculations, a brief review of calendar systems is helpful.

The biorhythm tables which follow are based on the characteristics of the calendar presently used in the United States (Georgian Calendar). Since calendars vary throughout the world, care should be taken when using dates of birth in foreign countries.

THE JULIAN CALENDAR

Julius Caesar put into use the first modern calendar in 45 B.C. He decreed that henceforth there shall be three years of 365 days each and then one year of 366 days in perpetual cycle. This calendar became known as the Julian Calendar. An inaccuracy is inherent in the Julian Calendar due to the fact that there are not exactly 365¼ days in a solar year, but 365 days, 12 hours, 48 minutes, 47.8 seconds. This annual difference of approximately eleven minutes becomes appreciable in the course of several centuries.

THE GEORGIAN CALENDAR

Pope Gregory XIII corrected the calendar in 1582. In order to make up for all the days which had accumulated since the beginning of the Julian Calendar, Pope Gregory XIII decreed the elimination of ten days from the year 1582, starting the day after October 4, 1582. The leap year rule, as used today, was also initiated. The Georgian leap year rule states that a leap year will not occur in a centesimal year (i.e., a year ending in 00) whose number cannot be divided by 400. This means that leap years do not occur in the years 1700, 1800, 1900. The leap year day is added to the last day in February. Since the year 2000 is divisible by 400, it will be a leap year and the month of February will have 29 days.

The American colonies switched to the Georgian Calendar in 1752 at the time the British Empire changed calendars. This resulted in September 2, 1752 being followed by September 14, 1752. Most dates in American history have been converted to the Georgian dates. The Sanhein Biorhythm Charts are based on the Georgian Calendar.

EVALUATING BIORHYTHMS WITHOUT COMPUTERS, PLOTTING OR EXTENSIVE CALCULATIONS

Sanhein developed a new and simplified method of computing biorhythms for publication in this book. The method involves the use of the Sanhein Biorhythm Charts (SBC) and allows the investigator to determine if a "hit" exists without computers, plotting, or extensive calculations. The SBC Method may be used to calculate the position of the biorhythm cycle for any date and birthdate, and is applicable to any accident analysis.

THE SBC METHOD

In the SBC Method, the biorhythms are considered to cross the axis at exactly midnight and the accident prone day exists for the next 24 hours, for days 1 and 15. For days 12 and 17 the biorhythm curve crosses the axis at 12 noon and the accident prone day exists for the next 24 hours ending at 12 noon on days 13 and 17 respectively.

CYCLE	*PHYSICAL (23 DAYS)*	*SENSITIVITY (28 DAYS)*	*INTELLECTUAL (33 DAYS)*
Accident Prone Days	1, 12, 13	1, 15	1, 17, 18

For discussion of accident prone days refer to Chapter 3.

To find if the accident prone day of the biorhythm cycle corresponds to the date of the accident Tables 8.1, 8.2 and 8.3 are used.

Table 8.1 is used to index an accident date or date of birth to the position of each of the cycles occurring on that day. For example on May 12, the cycle position is found by finding the intersection of the number 12 in the left-hand column and column 5. The block corresponding to this intersection is $\frac{132}{17\text{–}20\text{–}00}$. The top number is the corresponding day of the year which will be referred to as the "G" number. The three bottom numbers indicate the position of the physical, sensitivity and intellectual cycles respectively, corresponding to the 132 day of the year.

Table 8.1 Date Calendar With Biorhythm Cycles Imposed

DAY	#1 JAN	#2 FEB	#3 MAR	#4 APR	#5 MAY	#6 JUN
1	001 01-01-01	032 09-04-32	060 14-04-27	091 22-07-25	121 06-09-22	152 14-12-20
2	002 02-02-02	033 10-05-00	061 15-05-28	092 00-08-26	122 07-10-23	153 15-13-21
3	003 03-03-03	034 11-06-01	062 16-06-29	093 01-09-27	123 08-11-24	154 16-14-22
4	004 04-04-04	035 12-07-02	063 17-07-30	094 02-10-28	124 09-12-25	155 17-15-23
5	005 05-05-05	036 13-08-03	064 18-08-31	095 03-11-29	125 10-13-26	156 18-16-24
6	006 06-06-06	037 14-09-04	065 19-09-32	096 04-12-30	126 11-14-27	157 19-17-25
7	007 07-07-07	038 15-10-05	066 20-10-00	097 05-13-31	127 12-15-28	158 20-18-26
8	008 08-08-08	039 16-11-06	067 21-11-01	098 06-14-32	128 13-16-29	159 21-19-27
9	009 09-09-09	040 17-12-07	068 22-12-02	099 07-15-00	129 14-17-30	160 22-20-29
10	010 10-10-10	041 18-13-08	069 00-13-03	100 08-16-01	130 15-18-31	161 00-21-29
11	011 11-11-11	042 19-14-09	070 01-14-04	101 09-17-02	131 16-19-32	162 01-22-30
12	012 12-12-12	043 20-15-10	071 02-15-05	102 10-18-03	132 17-20-00	163 02-23-31
13	013 13-13-13	044 21-16-11	072 03-16-06	103 11-19-04	133 18-21-01	164 03-24-32
14	014 14-14-14	045 22-17-12	073 04-17-07	104 12-20-05	134 19-22-02	165 04-25-00
15	015 15-15-15	046 00-18-13	074 05-18-08	105 13-21-06	135 20-23-03	166 05-26-01
16	016 16-16-16	047 01-19-14	075 06-19-09	106 14-22-07	136 21-24-04	167 06-27-02
17	017 17-17-17	048 02-20-15	076 07-20-10	107 15-23-08	137 22-25-05	168 07-00-03
18	018 18-18-18	049 03-21-16	077 08-21-11	108 16-24-09	138 00-26-06	169 08-01-04
19	019 19-19-19	050 04-22-17	078 09-22-12	109 17-25-10	139 01-27-07	170 09-02-05
20	020 20-20-20	051 05-23-18	079 10-23-13	110 18-26-11	140 02-00-08	171 10-03-06
21	021 21-21-21	052 06-24-19	080 11-24-14	111 19-27-12	141 03-01-09	172 11-04-07
22	022 22-22-22	053 07-25-20	081 12-25-15	112 20-00-13	142 04-02-10	173 12-05-08
23	023 00-23-23	054 08-26-21	082 13-26-16	113 21-01-14	143 05-03-11	174 13-06-09
24	024 01-24-24	055 09-27-22	083 14-27-17	114 22-02-15	144 06-04-12	175 14-07-10
25	025 02-25-25	056 10-00-23	084 15-00-18	115 00-03-16	145 07-05-13	176 15-08-11
26	026 03-26-26	057 11-01-24	085 16-01-19	116 01-04-17	146 08-06-14	177 16-09-12
27	027 04-27-27	058 12-02-25	086 17-02-20	117 02-05-18	147 09-07-15	178 17-10-13
28	028 05-28-28	059 13-03-26	087 18-03-21	118 03-06-19	148 10-08-16	179 18-11-14
29	029 06-01-29		088 19-04-22	119 04-07-20	149 11-09-17	180 19-12-15
30	030 07-02-30		089 20-05-23	120 05-08-21	150 12-10-18	181 20-13-16
31	031 08-03-31		090 21-06-24		151 13-11-19	

NOTE: TOP NUMBER IS "G" DAY NUMBER
BOTTOM NUMBER IS BIORHYTHM CYCLE WITH PHYSICAL, SENSITIVITY AND INTELLECTUAL CYCLES DEPICTED FROM LEFT TO RIGHT.

Table 8.1 (Continued)

DAY	#7 JUL	#8 AUG	#9 SEPT	#10 OCT	#11 NOV	#12 DEC
1	182 21-14-17	213 06-17-15	244 14-20-13	274 21-22-10	305 06-25-08	335 13-27-05
2	183 22-15-18	214 07-18-16	245 15-21-14	275 22-23-11	306 07-26-09	336 14-00-06
3	184 00-16-19	215 08-19-17	246 16-22-15	276 00-24-12	307 08-27-10	337 15-01-07
4	185 01-17-20	216 09-20-18	247 17-23-16	277 01-25-13	308 09-00-11	338 16-02-08
5	186 02-18-21	217 10-21-19	248 18-24-17	278 02-26-19	309 10-01-12	339 17-03-09
6	187 03-19-22	218 11-22-20	249 19-25-18	279 03-27-15	310 11-02-13	340 18-04-10
7	188 04-20-23	219 12-23-21	250 20-26-29	280 04-00-16	311 12-03-14	341 19-05-11
8	189 05-21-24	220 13-24-22	251 21-27-20	281 05-01-17	312 13-04-15	342 20-06-12
9	190 06-22-25	221 14-25-23	252 22-00-21	282 06-02-18	313 14-05-16	343 21-07-13
10	191 07-23-26	222 15-26-24	253 00-01-22	283 07-03-19	314 15-06-17	344 22-08-14
11	192 08-24-27	223 16-27-25	254 01-02-23	284 08-04-20	315 16-07-18	345 00-09-15
12	193 09-25-28	224 17-00-26	255 02-03-24	285 09-05-21	316 17-08-19	346 01-10-16
13	194 10-26-29	225 18-01-27	256 03-04-25	286 10-06-22	317 18-09-20	347 02-11-17
14	195 11-27-30	226 19-02-28	257 04-05-26	287 11-07-23	318 19-10-21	348 03-12-18
15	196 12-00-31	227 20-02-29	258 05-06-27	288 12-08-27	319 20-11-22	349 04-13-19
16	197 13-01-32	228 21-04-30	259 06-07-28	289 13-09-25	320 21-12-23	350 05-14-20
17	198 14-02-00	229 22-05-31	260 07-08-29	290 14-10-26	321 22-13-24	351 06-15-21
18	199 15-03-01	230 00-06-32	261 08-09-30	291 15-11-27	322 00-14-25	352 07-16-22
19	200 16-04-02	231 01-07-00	262 09-10-31	292 16-12-28	323 01-15-26	353 08-17-23
20	201 17-05-03	232 02-08-01	263 10-11-32	293 17-13-29	324 02-16-27	354 09-18-24
21	202 18-06-04	233 03-09-02	264 11-12-00	294 18-14-30	325 03-17-28	355 10-19-25
22	203 19-07-05	234 04-10-03	265 12-13-01	295 19-15-31	326 04-18-29	356 11-20-26
23	204 20-08-06	235 05-11-04	266 13-14-02	296 20-16-32	327 05-19-30	357 12-21-27
24	205 21-09-07	236 06-12-05	267 14-15-03	297 21-17-00	328 06-20-31	358 13-22-28
25	206 22-10-08	237 07-13-06	268 15-16-04	298 22-18-01	329 07-21-32	359 14-23-29
26	207 00-11-09	238 08-14-07	269 16-17-05	299 00-19-02	330 08-22-00	360 15-24-30
27	208 01-12-10	239 09-15-08	270 17-18-06	300 01-20-03	331 09-23-01	361 16-25-31
28	209 02-13-11	240 10-16-09	271 18-19-07	301 02-21-04	332 10-24-02	362 17-26-32
29	210 03-14-12	241 11-17-10	272 19-20-08	302 03-22-05	333 11-25-03	363 18-27-00
30	211 04-15-13	242 12-18-11	273 20-21-09	303 04-23-06	334 12-26-04	364 19-00-01
31	212 05-16-14	243 13-19-12		304 05-24-07		365 20-01-02

Table 8.2

AGE IN YEARS	CYCLES P S I	AGE IN YEARS	CYCLES P S I	AGE IN YEARS	CYCLES P S I	AGE IN YEARS	CYCLES P S I
1	20-01-02	26	14-26-19	51	08-23-03	76	02-20-20
2	17-02-04	27	11-27-21	52	05-24-05	77	22-21-22
3	14-03-06	28	08-00-23	53	02-25-07	78	19-22-24
4	11-04-08	29	05-01-25	54	22-26-09	79	16-23-26
5	08-05-10	30	02-02-27	55	19-27-11	80	13-24-28
6	05-06-12	31	22-03-29	56	16-00-13	81	10-25-30
7	02-07-14	32	19-04-31	57	13-01-15	82	07-26-32
8	22-08-16	33	16-05-00	58	10-02-17	83	04-27-01
9	19-09-18	34	13-06-02	59	07-03-19	84	01-00-03
10	16-10-20	35	10-07-04	60	04-04-21	85	21-01-05
11	13-11-22	36	07-08-06	61	01-05-23	86	18-02-07
12	10-12-24	37	04-09-08	62	21-06-25	87	15-03-09
13	07-13-26	38	01-10-10	63	18-07-27	88	12-04-11
14	04-14-28	39	21-11-12	64	15-08-29	89	09-05-13
15	01-15-30	40	18-12-14	65	12-09-31	90	06-06-15
16	21-16-32	41	15-13-16	66	09-10-00	91	03-07-17
17	18-17-01	42	12-14-18	67	06-11-02	92	00-08-09
18	15-18-03	43	09-15-20	68	03-12-04	93	20-09-21
19	12-19-05	44	06-16-22	69	00-13-06	94	17-10-23
20	09-20-07	45	03-17-24	70	20-14-08	95	14-11-25
21	06-21-09	46	00-18-26	71	17-15-10	96	11-12-27
22	03-22-11	47	20-19-28	72	14-16-12	97	08-13-29
23	00-23-13	48	17-20-30	73	11-17-14	98	05-14-31
24	20-24-15	49	14-21-32	74	08-18-16	99	02-15-00
25	17-25-17	50	11-22-01	75	05-19-18	100	22-16-02

To find the position of the biorhythm cycles corresponding to the date of birth (G_2) and the date of the accident (G_1) an equivalent day (G_3) must be computed. Once the equivalent day G_3 is computed the corresponding position of the biorhythm cycles is found from Table 8.1. Equations 8-1 and 8-2 are used to compute G_3.

$$G_3 = G_1 - G_2 + 1 \qquad (8\text{-}1)$$

where: G_3 is the equivalent Biorhythm Day
G_1 is the day of the year corresponding to the accident date (Table 8.1)
G_2 is the day of the year corresponding to the birth date (Table 8.1)

If the birth is an earlier date (excluding year) than the accident the following relationship exists:

$$G_3 = 366 - G_2 + G_1 \qquad (8\text{-}2)$$

where G_1, G_2, and G_3 are as defined above.

The cycle positions corresponding to the Biorhythm Equivalent Day (G_3) are found directly below (G_3) in Table 8.1.

The position of the cycles corresponding to the year of the accident can then be found by adding in the cycle position, based on age at time of accident from Table 8.2 and the number of leap years which have occurred during that period from Table 8.3. People born in a leap year January through February 29 must add a day for that year. If the cycle numbers exceed 23, 28, or 33 days respectively, the number of completed cycles is subtracted.

Table 8.3 Leap Years

04,	08,	12,	16,	20,	24,	28,	32,	36,	50,	44,	48,
52,	56,	60,	64,	68,	72,	76,	80,	84,	88,	92,	96

NOTE: Leap years (February 29) occur every fourth year, but only those centesimal years (ending in "00") divisible by 400. A leap year did not occur in 1800 or 1900.

Use of the SBC Method is demonstrated in the following example.

SIMULATION 8-1

An accident occurred on September 9, 1976 involving a person born on May 12, 1936. Was this day an accident prone day in the biorhythm cycle?

ANALYSIS

The date of birth, May 12, is earlier than the accident date, thus Equation 8-1 should be used. Utilizing the values of G_1 and G_2 from Table 8.1,

$$G_3 = G_1 - G_2 + 1 = 252 - 132 + 1 = 121$$

The position of the cycles are calculated as follows:

		P	S	I
(a)	Cycle for day 121, from Table 8.1	6	9	22
(b)	Cycles for age 40, from Table 8.2	18	12	14
(c)	Number of leap years, from Table 8.3	10	10	10
(d)	*SUM*	34	31	46
(e)	Subtract completed cycles	23	28	33
(f)	Day in cycle	11	03	13

Thus, the biorhythm position is the 11th day in the physical cycle, the 3rd day in the sensitivity cycle and the 13th day in the intellectual cycle. These days do not correspond to accident prone days.

THE BIORHYTHM WORKSHEET

The Biorhythm Worksheet presented in Figure 8.1 may be used to summarize accident data analysis, and is useful in an in-plant study of accidents. In order to accurately determine a "hit," the exact birthdate and date and time of the accident must be known. Only accidents which occurred between 12 noon of days 12 and 17, and 12 noon of the following day should be included in the analysis. If the time data is not avail-

	DATE ACCIDENT	DATE BIRTH	(NOTE 1) CATEGORY 1A, B, C	(NOTES 2 & 3) G_3 Table 8.1			AGE (Y) Table 8.2			LEAP YEAR (L) Table 8.3	COMPOSITE $G_3 + Y - L$			(NOTE 4) HITS P S I		
NO.	G_1 Table 8.1	G_2 Table 8.1	II	P	S	I	P	S	I		23 P	28 S	33 I	1 12 13	1 15	1 17 18

NOTES:

1. ACCIDENT CATEGORY
   I. LOST TIME ACCIDENT (24 HRS)
   II. NON LOST TIME ACCIDENT (LESS THAN 24 HRS)
   A. INDIVIDUAL PERFORMANCE
   B. THIRD PARTY
   C. BEYOND INDIVIDUAL'S CONTROL
2. $G_3 = G_1 - G_2 + 1 \ (G_2 < G_1)$
3. $G_3 = 366 - G_2 + G_1 \ (G_1 < G_2)$
4. HITS - Y (YES), N (NO)

Figure 8.1 Biorhythm Accident Worksheet

able, analysis may be made on a 48-hour basis, however the random probability of hits will increase from 20% to 26% (refer to Table 2.2, Chapter 2).

SIMULATION 8-2

A plant had a total of eleven accidents in 1975. The Log of Occupational Injuries and Illnesses is presented in Figure 8.2. Test the accidents for a possible relationship with the biorhythm cycles.

ANALYSIS

Step 1 The first step is to determine which accidents to exclude from the analysis. From Figure 8.2 it is seen that cases number 4 and 5 are skin rashes and should not be considered. Case 7 involves intoxication and is also excluded. The remaining eight cases involve accidents which result from human performance and will be considered.

Step 2 Next, the birth date for each accident victim and the time of the accident are compiled. This information is entered in the worksheet, as illustrated in Figure 8.3.

Step 3 The worksheet (Figure 8.3) is completed and the appropriate calculations are made. Of the eight accidents, four occurred on accident prone days with a resulting "hit" rate of 50%. Since the sample size is small the application of the Goodness Fit Test is not meaningful, confidence limits are used. Based on a 90% confidence limit the true mean is expected to lie between $p \pm a\ s$

Where $p = 4/8 = .5$

$$S = \sqrt{\frac{.5(.5)}{8}} = .176 \qquad (2\text{-}8)$$

From Table 2.4, a = .63. Thus with a 90% confidence limit the proportion of hits could be as low as .5 − .83 (.176) = .35.

Since the expected proportion of hits is .20, the results of this study could not have occurred by pure chance alone.

OSHA NO. 100

LOG OF OCCUPATIONAL INJURIES AND ILLNESSES

RECORDABLE CASES: You are required to record information about every occupational death; every nonfatal occupational illness; and those nonfatal occupational injuries which involve one or more of the following: loss of consciousness, restriction of work or motion, transfer to another job, or medical treatment (other than first aid). More complete definitions appear on the other side of this form.

Form Approved OMB No. 44R 1453

CASE OR FILE NUMBER	DATE OF INJURY OR ONSET OF ILLNESS	EMPLOYEE'S NAME (First name or initial, middle initial, last name)	OCCUPATION (Enter regular job title, not activity employee was performing when injured or at onset of illness.)	DEPARTMENT (Enter department in which the employee is regularly employed.)	DESCRIPTION OF INJURY OR ILLNESS: Nature of Injury or Illness and Part(s) of Body Affected (Typical entries for this column might be: Amputation of 1st joint right forefinger; Strain of lower back; Contact dermatitis on both hands; Electrocution—body)	DESCRIPTION OF INJURY OR ILLNESS: Injury or Illness Code. See codes at bottom of page	EXTENT OF AND OUTCOME OF CASES: DEATHS (Enter date of death.)	LOST WORKDAY CASES: Enter a check if case involved lost workdays.	LOST WORKDAYS: Enter number of days AWAY FROM WORK due to injury or illness.	LOST WORKDAYS: Enter number of days of RESTRICTED WORK ACTIVITY due to injury or illness.	NONFATAL CASES WITHOUT LOST WORKDAYS (Enter a check if no entry was made in columns 8 or 9 but the case is recordable, as defined above.)	TIME OF ACCIDENT
(1)	No./day/yr. (2)	(3)	(4)	(5)	(6)	(7)	Mo./day/yr. (8)	(9)	(9A)	(9B)	(10)	(11)
1	1/18/75	JOE STRAIN	HELPER	CHEMICAL	STRAINED BACK	10		✓	1			10 AM
2	3/13/75	PHIL CUT	CLERK	ADMIN	CUT HAND	10		✓	1			4 P.M.
3	1/29/75	JOHN BRUISE	MECHANIC	MAINTENANCE	BRUISED MOUTH	10		✓	2			1 P.M.
4	3/19/75	AL SKIN	HELPER	CHEMICAL	SKIN RASH	21		✓	1			6 A.M
5	4/7/75	BOB RASH	HELPER	CHEMICAL	SKIN RASH	21		✓	1			3 P.M
6	4/22/75	TONY FOOT	OPERATOR	PRODUCTION	BRUISED LEFT FOOT	10		✓	1			5 P.M
7	1/29/75	BILL ILL	OPERATOR	PRODUCTION	ST. INTOXICATION	25		✓	2			2 P.M.
8	3/13/75	JAKE SHOULDER	SHIPPING CLERK	ADDITIVES	CONTUSION RT. SHOULDER	10		✓	1			4 P.M.
9	6/2/75	HANK MASHER	HELPER	CLERICAL	MASHED FINGER	10		✓	5			3 P.M.
10	7/9/75	AL TOE	MECHANIC	MAINTENANCE	BRUISED RT. BIG TOE	10		✓	1			11 A.M
11	6/3/75	WRIGHT EYE	CLERK	ADMIN	BRUISED RT EYE	10		✓	1			2 P.M

Company Name ________
Establishment Name ________
Establishment Address ________

NOTE This is NOT a report form. Keep it in the establishment for 5 years.

Injury Code
10 All occupational injuries
Illness Codes
21 Occupational skin diseases or disorders
22 Dust diseases of the lungs (pneumoconioses)
23 Respiratory conditions due to toxic agents
24 Poisoning (systemic effects of toxic materials)
25 Disorders due to physical agents (other than toxic materials)
26 Disorders associated with repeated trauma
29 All other occupational illnesses

Figure 8.2 Log of Occupational Injuries and Illnesses

NO.	DATE ACCIDENT G_1 Table 8.1	DATE BIRTH G_2 Table 8.1	(NOTE 1) CATEGORY 1A, B, C II	(NOTES 2 & 3) G_3 Table 8.1 P	S	I	AGE (Y) Table 8.2 P	S	I	LEAP YEAR (L) Table 8.3	COMPOSITE $G_3 + Y + L$ 23 P	28 S	33 I	(NOTE 4) HITS P 1 12 13	S 1 15	I 1 17 18
1	18	213	IA	171			24			6	13	5	27	Y	N	N
	1-18-75	8-1-50		10	3	6	20	24	15							
2	72	100	IA	338			30			7	2	11	9	N	N	N
	3-13-75	4-10-44		16	2	8	2	2	27							
3	29	255	IA	140			48			12	8	4	17	N	N	Y
	1-29-75	9-12-76		2	0	8	17	20	30							
4			IC													
	3-19-75															
5			IC													
	4-7-75															
6	112	69	IA	44			22			5	6	15	27	N	Y	N
	4-22-75	3-10-53		21	16	11	3	22	11							
7			IC													
	1-29-75															
8	72	100	IA	338			30			7	2	11	9	N	N	N
	3-13-75	4-1044		16	2	8	2	2	27							
9	190	139	IA	52			27			6	23	1	13	N	Y	N
	7-9-75	5-19-48		6	24	19	11	27	21							
10	153	228	IA	291			61			15	8	3	32	N	N	N
	6-2-75	8-16-13		15	11	27	1	5	23							
11	154	67	IA	88			21			5	7	2	3	N	N	N
	6-3-75	3-8-54		19	4	22	6	21	9							

Figure 8.3 Accident Worksheet for Simulation 8-2

NOTES:

1. ACCIDENT CATEGORY
   - I. LOST TIME ACCIDENT (24 HRS)
   - II. NON LOST TIME ACCIDENT (LESS THAN 24 HRS)
   - A. INDIVIDUAL PERFORMANCE
   - B. THIRD PARTY
   - C. BEYOND INDIVIDUAL'S CONTROL
2. $G_3 = G_1 - G_2 + 1\ (G_2 < G_1)$
3. $G_3 = 366 - G_2 + G_1\ (G_1 < G_2)$
4. HITS - Y (YES), N (NO)

SIMULATION 8-3

Assume the results of the analysis presented in Simulation 8-2 are repeated ten identical times in the analysis of 110 accidents. For this analysis, determine if the results are statistically significant at a 90% confidence level.

ANALYSIS

This sample size is adequate for using the Goodness Fit Test. The probabilities of Table 2.2 are used to compute the expected number:

Accident Prone Days	*Observed Number*	*Expected Number*	$\frac{(n_i - np_i)^2}{np}$
Non-Accident Prone Days	40	.79 X 80 = 63	8.3
P	10	6	2.6
S	20	4.9	4.5
I	10	4.1	8.4
P X S	0	.46	.46
P X I	0	.39	.39
E X I	0	.31	.31
P X E X I	0	.031	.031
		λ^2 =	24.991

From Table 2.3 corresponding to P = .1, λ^2 = 12.02 (with 7 degrees of freedom). Since 24.9 > 12.09, the results should not be attributed to pure chance.

A BIORHYTHM EXPERIMENT

If the results of the initial analysis indicate a positive correlation between accidents and the biorhythm cycles, an experiment may be performed to determine the potential effectiveness of a biorhythm program as a method to prevent accidents. This experiment involves dividing employees into three groups:

Group I – Control Group – no biorhythm information is distributed to this group.

Group II – Bio-Group – Correct biorhythm information is distributed to this group indicating projected accident prone days.

Group III – Neg-Bio-Group – Incorrect biorhythm information is distributed to this group, indicating incorrect accident prone days.

From the results of this experiment the effect of safety awareness on a random basis (Group III) and the additional effect of the knowledge of the individual's own biorhythm accident prone days (Group II) may be determined. If a significant decrease in accident occurrences is found, a plant may choose to adopt a biorhythm program on a permanent basis.

USING BIORHYTHMS TO PREVENT ACCIDENTS

Biorhythm theory may be applied to reduce the frequency of industrial accidents by informing employees of their accident prone days. When employees are aware of accident prone periods, they may exert extra caution to prevent possible accidents. Several methods of informing individuals of their accident prone days may be employed:

1. An employee's time card may list the individual's accident prone days for a week.
2. Biorhythm information may be listed on an employee's paycheck.
3. Time cards may be coded with employees' birthdates. The time clock may be programmed to warn the individuals of accident prone days as they punch in.
4. Employees may be provided with biorhythm tables, and those who choose to participate in the biorhythm program may compute their own biorhythms.
5. Supervisors may be informed on a daily basis of the accident prone days for the employees in their de-

partment, and in turn, personally inform workers to exert appropriate caution. For occupations involving unusual hazards, an employee may be shifted to a less critical task on "double" or "triple" accident prone days.

6. Employees may be issued a red-colored hard hat or badge on accident prone days to provide a constant reminder to both the individual and his co-workers.

PROJECTING ACCIDENT PRONE DAYS

The biorhythm accident prone days can be projected by using the SBC Method, Tables 8.1, 8.2, and 8.3.

SIMULATION 8-4

Bernie Random was born on March 8, 1954. What are Bernie's accident prone days for the month of January 1977?

ANALYSIS

NO.	FORECAST DATE	DATE BIRTH	G_3 Table 8.1			AGE (Y) Table 8.2			LEAP YEAR (L) Table 8.3	COMPOSITE $G_3 + Y + L$		
	G_1 Table 8.1	G_2 Table 8.1	P	S	I	P	S	I		23 P	28 S	33 I
1	1	67	300			22			6	10	20	20
	1-1-76	3-8-54	1	20	3	3	22	11				

Thus for the month of

JANUARY	23 P	28 S	33 I	JANUARY	23 P	28 S	33 I
1	10	20	20	17	3	8	3
2	11	21	21	18	4	9	4
3	12*	22	22	19	5	10	5
4	13*	23	23	20	6	11	6
5	14	24	24	21	7	12	7
6	15	25	25	22	8	13	8
7	16	26	26	23	9	14	9
8	17	27	27	24	10	15*	10
9	18	28	28	25	11	16	11
10	19	1*	29	26	12*	17	12
11	20	2	30	27	13*	18	13
12	21	3	31	28	14	19	14
13	22	4	32	29	15	20	15
14	23	5	33	30	16	21	16
15	1*	6	1*	31	17	22	17*
16	2	7	2				

* ACCIDENT PRONE DAYS — JANUARY 3, 4, 10, 15, 24, 26, 27, 31

USE OF PLOTS

An alternate method to project accident prone days involves overlaying the physical, intellectual, and sensitivity cycles on monthly graph paper. Figures 8.4, 8.5, 8.6 may be used as templates for this purpose. A color code for curves suggested by Thommen is that red is used for the physical rhythm, blue for the sensitivity rhythm and green for the intellectual rhythm.

The curves indicated in Figures 8.4, 8.5 and 8.6 should be cut out carefully and pasted on lightweight cardboard. To draw the curves, the position of the curve is lined up with the calculated position of the biorhythm cycle and the curve is traced with a pencil.

SIMULATION 8-5

For Simulation 8-4 draw the three biorhythm curves.

ANALYSIS

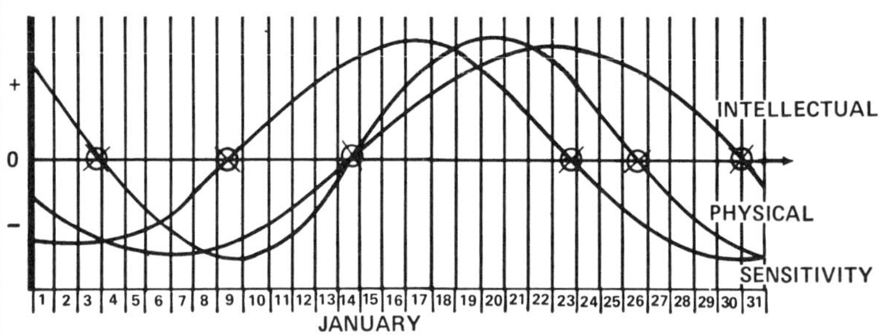

As indicated in Simulation 8-4 on January 1, the physical cycle is in its 10th day, the sensitivity cycle is in its 20th day and the intellectual cycle is in its 20th day. The three biorhythm curves should be carefully lined up with the corresponding calendar date.

CUT OUT CURVE
AND PASTE ON
CARDBOARD

*Use GREEN Color
to Plot*

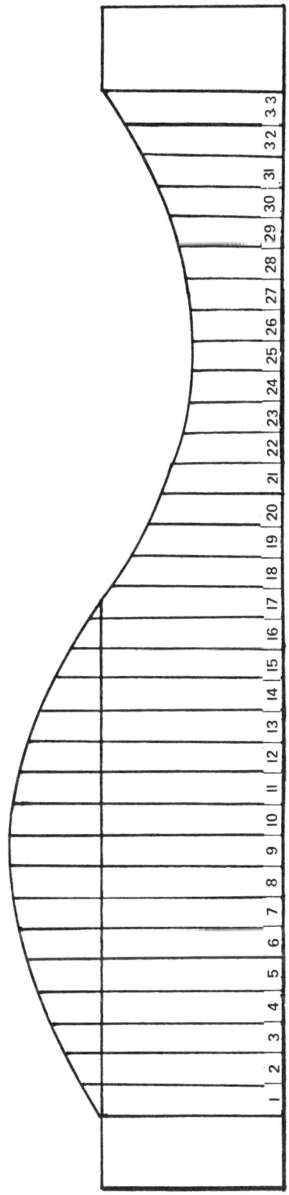

Figure 8.4 Intellectual Cycle

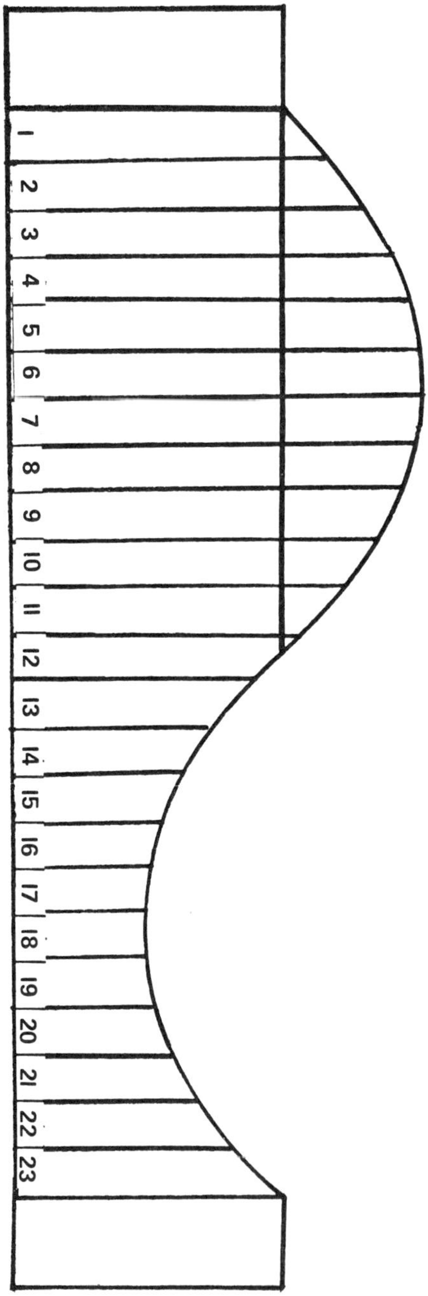

CUT OUT CURVE
AND PASTE ON
CARDBOARD

*Use RED Color
to Plot*

Figure 8.5 Physical Cycle

CUT OUT CURVE
AND PASTE ON
CARDBOARD

*Use BLUE Color
to Plot*

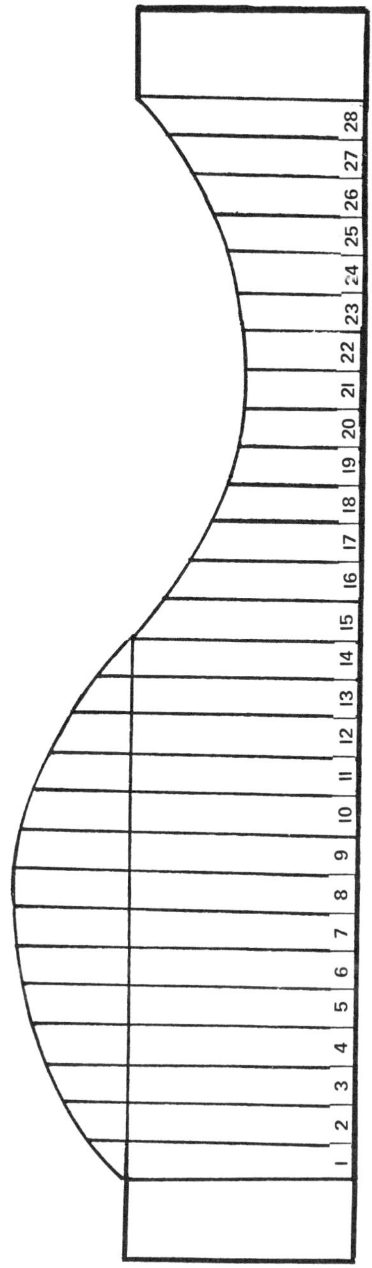

Figure 8.6 Sensitivity Cycle

BIORHYTHM PRODUCTS AND SERVICES

There are many products and services currently on the market which are based on the biorhythm theory. The summary which follows is neither a complete representation nor an endorsement, but simply serves as a representative sample.

BIORHYTHM HAND CALCULATORS

The calculations of the SBC Method may also be accomplished with a biorhythm hand calculator. One such calculator, the Casio Biolator H-801 Electronic Calculator, is shown in Figure 8.7.

Figure 8.7 Biorhythm Hand Calculator

CYCLOGRAF KIT

The Cyclograf Kit contains sine curve rulers and preprinted charts.

DIALGRAF

The Dialgraf is an easy to use device which gives relative positions of biorhythms on any given day.

Cyclograf Kit and Dialgraf are available from Edmund Scientific Co., 150 East Gloucester Pike, Barrington, N. J. 08007. Total cost approximately $15.00.

These items are also available from ARCO Publishing Company, Inc., 219 Park Avenue South, New York, N. Y. 10003.

BIO-KIT

The Bio-Kit contains calibrated sine curves and preprinted charts. Suggested list price is $4.95. This kit is available from PSI Rhythms, Inc., 2382 South Dixie Avenue, Dayton, Ohio 45409.

PERSONAL COMPUTER BIORHYTHM CHARTS

These charts illustrate an individual's biorhythms and give a subjective rating as to probable responses on each calendar day. The cost of these services range between $5.00 and $49.95 per individual. Figure 8.8 illustrates a typical output.

These services are available from:

BIOCOM 300
Bio-Safety Consultants, Inc.
2625 West Britton Road
Oklahoma City, Oklahoma 73120

Biorhythm Barometer Charts
P. O. Box 80309
Atlanta, Georgia 30366

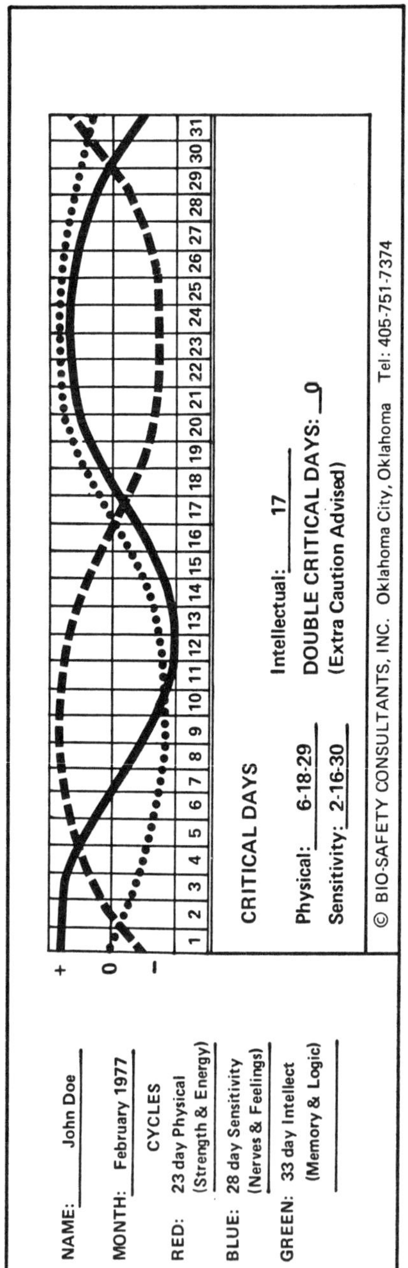

Figure 8.8 Typical Computer Readout

Bio-Rhythm Chart Report
Time Pattern Research Inc.
21 Henderson Drive
West Caldwell, N. J. 07006
Also available through Edmund Scientific Company

Bio-Chart
ARCO Publishing Company, Inc.

BIORHYTHM TABLES

In addition to the biorhythm tables incorporated in this book two other commonly used ones are available.

George S. Thommen Tables
Is This Your Day?
Crown Publishers, Inc., New York
(160 pages)

Bernard Gittelson Tables
Biorhythm—A Personal Science
ARCO Publishing Company, Inc.
New York
(186 pages)

Eighty-three pages are devoted to the Gittelson Tables. The reason the tables are lengthy is that for each individual's birthdate a letter designation is given. By drawing a line through these letter designations the biorhythm curves are drawn for the years 1976 through 1978.

RELATED BOOKS

Other related books include:

Biorhythms In Your Life
Daniel Cohen
Fawcett Publications Inc., 1976

Body Time
Gay Gaer Luce
Bantam Books, Inc., 1971

UTILIZING OUTSIDE CONSULTANTS

The options facing the employer instituting a program is whether to create an internal department charged with the responsibility of gathering and disseminating biorhythmic information or to engage a consulting firm specializing in this phase of business operations.

A plethora of literature is now available that should enable an industry or business to establish the mechanical means necessary to gather meaningful information which may be used in conjunction with the previously mentioned computer programs. Company employees assigned the job of analysis and distribution must have thorough training in the uses and values of biorhythmic information, and, if the computer is employed, in its operation.

The experience of some industries utilizing "in-house" personnel to staff a biorhythm analysis department has shown that there is a tendency for the employees to regard their duties in gathering biorhythmic data as a secondary task; subsequently the information is often sketchy and it is prepared and disseminated in a haphazard manner.

For these, and other reasons, many industries engage consulting firms to establish and operate their biorhythmic oriented safety program. One such firm currently active in this area is Bio-Safety Consultants, Incorporated of Oklahoma City. Executives of that company explain that their primary advantage is the ability to establish an "overview" of a client's operations. Because personnel of Bio-Safety are not involved in the daily operations, and since biorhythmic analysis and applications of gathered data is their only interest, they are often able to detect and point out potentially hazardous or wasteful situations before they become critical.

As is the case with their counterparts in Japan, some of Bio-Safety's major clients are in the transportation industry. In a typical situation, personnel of the consulting firm will initiate service by examining and evaluating the records of the company employing their services. Accident reports are scrutinized and victim's cycles are prepared in graphic form to determine a correlation or a lack of correlation between critical days and biorhythmic profiles. This information is prepared and processed through the use of the Biocom 300 computer, and statistical correlations are made on an IBM 370 computer.

After the original evaluation is completed, the consulting firm makes recommendations to management relating to ways in which biorhythmic information can best be utilized in the client corporation's operations and continues to supply profiles, information, and advice through the agreed period of contractual obligation.

DISCUSSION

Analysis of the research studies of Sanhein, Wolcott, and Neil may raise many questions in this new science.

Sanhein's study, as presented in Chapter 3, seems to clearly indicate a correlation between accidents and the biorhythm cycles. However, one unanswered question is whether the accidents discarded from the study have tended to distort the conclusions. Although accidents which are not human factor related should be eliminated from the analysis, their elimination is a subjective process unless guidelines are established. (Guidelines for accident selection have been presented in this chapter.) In Sanhein's study, a total of 217 accidents were discarded from the Naval Weapons Center Study. By random chance, the number of "hits" on accident prone days for the data discarded should correspond to that predicted by normal chance. An analysis of the data presented indicates that the number of "hits" for the discarded data is below that expected by random chance. This raises a question as to the degree of certainty that the data used reflects the total population. At best, biorhythm cycles are only one of several influences on human performance which influence accidents where routine, repetitive tasks are involved.

While Sanhein found a correlation between biorhythms and industrial accidents, he found no correlation with suicides.

Wolcott's study indicates that there is no correlation with aircraft accidents and biorhythms. The question can be raised again as to whether in this study accidents which were not a result of human performance were included in the analysis. Inclusion of these accidents would tend to lower the "hit" percentage. The National Transportation Safety Board reviewed in detail each accident and determined which accidents were caused primarily by pilot error. Since this was the case it is felt that the accidents included in the report represented human performance errors. If Wolcott's results are correct, it proves that no correlation exists for aircraft accidents. It does not prove, however, that no correlation exists for other occupations. For pilot involved aircraft accidents to occur, several checks must be voided. A higher degree of sophistication of procedures to prevent these types of accidents are present, and this may in itself void the mild influence of the biorhythm cycles.

The results of Neil's experiment, presented in Chapter 7, indicate that there is periodicity in human performance which follows the biorhythm time periods. The question raised is whether these cycles start at the time of birth and therefore can be predicted. Further experimentation could verify whether the biorhythm cycles are initiated at birth.

The four studies presented in this text were chosen for their degree of sophistication and the methodology used. The questions raised in no way reflect the quality of the studies or the abilities of the researchers. They do however reflect on the state-of-the-art of the "industrial biorhythm science."

In an attempt to answer these questions, Dr. Douglas Neil was contacted. He indicated that even though his study did not mention correlation of the periods in the experiment with date of birth, this relationship was found to exist. The performance periods were compared to the biorhythm curves computed for the specified period. The relationship was apparent. Neil also indicated that he believes that challenging tasks bring out the positive influence of the biorhythm cycles. In other words the biorhythmic influence is not as easily masked by other factors during periods of challenge. He does not believe that the biorhythm model is perfect due to the influence of other factors. However, it does provide a simple model of human performance which does predict behavior better than any other model presently in practice.

The question raised about Sanhein's study can be approached from a different viewpoint. If random chance probabilities are applied to the accidents discarded, it is computed that a maximum of 112 additional noncritical accidents should have been included in the biorhythm analysis of the Navy Weapons Center Study (based on an expected probability of hits of 26.6%). By including 112 additional accidents in the experiment and maintaining the same number of hits (189), the percentage of hits is reduced from 42.37 to 33.81. Thus we can argue over the exact percent hit, but the results still conclusively prove a relationship with the biorhythm cycles. If a Chi Square statistical test is applied to the Sanhein data the null hypothesis of random chance is rejected, and thus a correlation is supported.

Certain industries find that there is a correlation with accidents and biorhythms while other industries find no correlation. The differences may lie in the nature of the work activity. If very large samples are utilized based on data from ***several*** industries, it is possible that no correlation would be found. Mason's study, referred to in Chapter 1, seems to confirm this.

SUMMARY

Biorhythms as applied to industrial safety is a controversial subject. The reason is partially due to unsubstantiated exaggerated reports and a lack of understanding of statistics. Any biorhythm study should contain the following:

1. Basis of excluding data
2. Definition of accident prone days and time period for including "hits"
3. Number of accidents studied
4. Accident time and birth date of employee
5. Statistical significance
   a. Goodness fit
   b. Confidence interval
   c. Other

This information will enable studies from various companies to be compared on the same basis. The rewards of proving or disproving biorhythms are appreciable. Even if only a small percentage of industrial accidents can be prevented, the value of biorhythms would be enormous.

The emphasis on accident prevention and the better understanding of human behavior will prove beneficial even if the results of further studies minimize the probable effects of biorhythms. Science has looked skeptically on biorhythms, and for good reason. The time is ripe for the science of biorhythms to be proven or disproven conclusively. Hopefully this book will give you the tools to "get started" on this important project.

Index